The Cube

Detailed Design Spec for a 3D Volumetric Solid State Display Device

Vesta Imprints

San Francisco

The Cube
Detailed Design Spec for a 3D Volumetric Solid State Display Device

Published by Vesta Imprints, San Francisco, California, USA

ISBN: 978-1-7347962-2-3

First printing August 2021
Round 1 edits September 2021
Round 2 edits November 2021

Public Domain cover artwork by Justin Maller

This book is based on technical papers published by the author
for the LinkedIn community between July 2014 and Nov 2019.

Table of Contents

Chapter 1

An RGB 3D Display Device Without Moving Parts

1.0. Introduction

The purpose of this book is to design on paper a solid-state 3D volumetric display device. *Solid state* means there are no moving parts. One patented design by Ed Berlin for a 3D display device uses a rotating armature to display the image. This has the disadvantage of moving parts and motor control, as well as the constraint of the volume of a central column through which all of the wiring is routed.

Tests performed in software using voxels indicate that a resolution of 128 cubed is the minimum size for a voxel image. Below this size the image becomes too blocky to be readily recognizable. As well, the physical size of the display device for this resolution is found to be 2 inches square. Above this size the eye doesn't readily fuse the individual voxels into a continuous image, and below this size the image is really small.

One theme of this book is nanotechnology and *metamaterials*. The technology of metamaterials is about two decades old; some simple 2D flat-plane metamaterial prototypes have been built and the theory has been validated. A metamaterial is a macroscopic material structure having a man-made one, two, or three dimensional cellular architecture. Metamaterials produce material parameters (permittivity, permeability, and index of refraction) that are not found in nature or even in conventional composite materials. The main feature of metamaterials is that, using a design computed with conformal mapping and electromagnetic equations, a covering can be manufactured which renders a physical cylinder invisible. Electrical signals can be conducted through wires which run interior to the cylinder, thus rendering them invisible. However there are many roadblocks and limitations to this idea and a practical implementation is still in the future.

Though the actual manufacture of "cloaked transmission lines" is not yet a reality, the pace of this work is such that it will be a reality within two decades, and so this book will look at a preliminary design for a 3D display device which makes use of metamaterial conductors that are optically invisible.

1.1. Software

In the rotating armature design, the basic voxel grid can't be displayed directly. It has to be further processed to accomplish a radial mapping that matches the rotating armature. This has the disadvantage of "sampling a sample," as well as overlap of samples on the inner parts of the arms, and the danger of undersampling on the outer parts.

For a stationary design based on a rectangular grid there is no need to sample the sample. The original voxel representation in software can be used directly in hardware.

1.2. Hardware

In the rotating armature design, complicated processing is required for triggering LEDs embedded in the radial arms at the proper moment. The bigger problem is real estate. Routing thousands of wires through the central column of the rotating armature is problematic.

As well, placing 60 LEDs on a radial arm less than one inch in length, wiring them, and still leaving space for being able to see the light emission is a big challenge (unmet). The wires, no matter how thin, no matter how many layers are used, no matter how coiled or spiraled, will block the light on each arm.

The metamaterial approach eliminates both of these problems. Instead of a rotating armature with a central column, the design calls for a base and a removable transparent cube that rests on top of it.

In the new scheme the placement of wires is still a tricky business. There still needs to be a minimum separation between cloaked wires because the physical wires are still there even though the cloaks bend the light around them. In other words, some intervening space is necessary for the LEDs behind the cloaked wires to be able to shine through the grid as seen from some viewing angle.

The scale of the invisibility in one prototype [2] corresponds to a wire diameter of about 10 micrometers. Given a two-inch display device, we start with 2 inches = 5.08 centimeters, or 50.8 millimeters. This is divided by 128, giving a basic trace width of 396 micrometers or *microns*. The percentage of that width needed for light emission is estimated to be 75, which leaves 25 percent or 100 micrometers for each conductor.

1.3. Solid Cube

The cube is nearly indestructable. The cube should fit in the palm of one's hand and one should be able to juggle it and toss it around without damage. It should appear mostly clear, with no hint of the millions of connections inside. When the cube is placed in its cradle, a full 3D image should appear instantly. But the cube is also equipped with batteries, so that it can be removed from the base temporarily and the image should persist.

When viewed from an angle looking down at three of the cube's six faces, e.g. the three faces closest to the viewer, the wide-angle view afforded by the three front-facing surfaces is more than adequate to give the viewer the perception that the image is three-dimensional and fills space.

Some applications would benefit from a full 360-degree view, such as surgery previews. On the other hand, an application such as a driver display wouldn't need it since it would be embedded in the windshield glass and the driver would always face the windshield from the driver's seat.

The cube manufacture would require the fusing of many thin layers which sandwich the traces and light emitters – without compromising the dielectric material that composes the cloaked transmission lines. Visual clarity between layers would have to be maintained. The high-temperature fusing of glass and silicon layers is a project in itself.

1.4. Cloaking Technicalities

In general there are two kinds of cloaked objects: *passive* and *active*. A *passive* object has no internal currents and is non-conducting. Examples include a point which can be surrounded by a cloaking sphere; or a line surrounded by a cloaking cylinder. An *active* object has internal currents which interfere with the mathematical transformation that underlies all physical cloak designs. An active object must be *double cloaked.*

A key topological fact is that the exterior N consists both of the outside of the cloaked cylinder, N_1, and the cylinder bore, N_2. This is shown below.

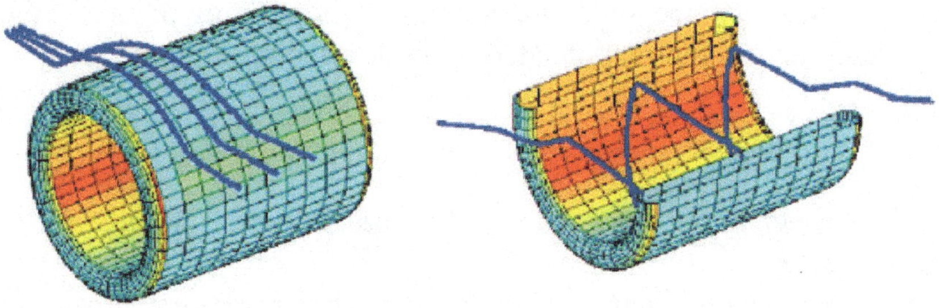

(courtesy [1])

Ideally, one is looking for *locally finite energy* solutions for all incoming waves. This can be accomplished if the material of the skin satisfies the soft-and-hard surface (SHS) boundary conditions for Maxwell's equations. These boundary conditions are outlined in [4]. Note that SHS conditions have not been achieved in the lab and are still theoretical, but Greenleaf notes in [1] that the boundary condition can be achieved with a perfectly conducting material with corrugations on its surface. This is rephrased in [4] as thin parallel gratings filled with a dielectric material – similar to the situation in [3], that is, a dielectric filled with a grating of elliptical silver nanowires.

The cloaking shell has dispersion properties which affect a lightwave trajectory, and the material gives a magnetic response in the optical range. In the case of a transverse magnetic (TM) incidence design, the magnetic field is polarized along the z-axis, that is, along the length of the cloaked transmission line. The key to the design is the resonance between the cells that make up the dielectric skin. This resonance depends strongly on frequency, which is why almost all metamaterials operate within a narrow bandwidth.

The authors simulated the design in [3] by using a finite-element mesh modeler in 2D. The design was tested with light shining in the transverse magnetic (TM) direction only, at the frequency of a helium-neon laser. The 3D case is more difficult because light will be arriving from all directions. If ambient light from any direction is too ambitious a goal for invisibility, it may be possible to manufacture a metamaterial with a preferred set of "grain" directions, i.e., with limited anisotropies, to achieve an invisibility effect within the full viewing angle of the display cube. Assuming that all of these technical challenges can be solved, then the wires that drive the display should be invisible.

Unfortunately the equations at the boundary of the cloaked surface "blow up" because the math becomes singular, and so the solution must be truncated. This affects the quality of the cloak. The area of mathematics that deals with shape-dependent electromagnetic responses is called Effective Medium Theory (EMT), first used in cosmology by H.C. van de Hulst to model light scattered by small spaceborne particles. EMT was used in [3] to compute the "screen factor" of nanowires embedded in the dielectric. The production of such metamaterials is still in the future but the software simulation performed in [3] is fully valid because there is a close tie between theoretical models and built reality – that is, if the models can be built in the first place.

1.5. Conclusion

This chapter has presented some of the major elements for the design of a 3D display cube. The display uses transmission lines which run vertically through the cube. Each transmission line is rendered invisible with the aid of a nanomaterial invisibility cloak that surrounds it. This design is superior to an older rotating armature design because there are no moving parts and no sampling requirements nor timing constraints based on rotation speeds. The armature design achieves transparency only through fast rotation: the arm is still visible as a blur. The display cube, by comparison, uses no moving parts and there would be no smearing or timing issues such as those associated with radial motion. The image would be rock-steady.

References

[1] Greenleaf, et al. *Electromagnetic Wormholes and Virtual Magnetic Monopoles from Metamaterials*, Nature, v. 450, November 2007
[2] Smolyaninov, et al. *Electromagnetic Cloaking in the Visible Frequency Range*, arxiv 0709.2862, 2007
[3] Cai et al. *Optical Cloaking with Metamaterials*, Nature Photonics, v. 1, April 2007
[4] Greenleaf, et al. *Cloaking Devices, Electromagnetic Wormholes and Transformation Optics*, SIAM Rev., v. ?, 2008
[5] Weder, Ricardo. *The Boundary Conditions for Point Transformed Electromagnetic Invisibility Cloaks*, arxiv 0801.3611 v.2, 2008

Chapter 2
Transmission Lines

2.0. Introduction
Chapter 1 envisioned a transformation optics mapping involving wormhole topology, first espoused by Greenleaf *et al* [1], to cloak signal transmissions lines in a dense 3D cube. Unfortunately no other papers except those by Greenleaf could be found that deal with this wormhole transformation.

Further research made clear that there are many limitations and outstanding research problems with the entire metamaterial genre, such as limited frequency, limited viewing angle, temperature dissipation, and so forth. Moreover, it became clear that the wormhole paradigm has limitations above and beyond the difficulty of the conformal mapping – the main one being a practical interface between the wormhole mouths and the external LED circuits.

2.1. Approaches to Transmission
The wormhole approach described by Greenleaf is a good conceptual paradigm but not a very practical one. The name "wormhole" conjures up images of collapsed stars but the astrophysical reality has hardly anything to do with transformation optics. There's no need to enumerate the difficulties of the wormhole transformation from a practical viewpoint.

Optical effects take place in the terahertz range, which means that metamaterials have to be designed at a very small scale, and these scales are hard to achieve. The 3D display cube can be characterized as a *densely packed matrix*. This poses a further challenge for the practical design of transmission lines. The following considerations are paramount:

a. The transmission lines must be cloaked.
b. The transmission lines must consume minimal real estate, i.e., they must be thin.
c. The transmission lines must transmit signals which drive the LEDs on each level.

The transmission line exterior should consist of metamaterials tuned to optical frequencies to cloak the interior. Zhang and Wu have pushed the envelope on metamaterial cloaking with their paper *Cylindrical Cloaking At Oblique Incidence With Optimized Finite Multilayer Parameters* [5].

The key result of this paper is that invisibility can be achieved over an "incidence of viewing" angle of 100 degrees (+/- 50 degrees from 0 degrees, the surface normal) by using multiple layers (in their simulation, four). The layer parameters are found through numerical optimization.

Since the typical viewing angle of the display cube is within the 50 degree limit, under these conditions the transmission lines extending vertically through the color planes would be cloaked. However looking at the cube from a higher viewing angle would reveal the cloaks.

The design for the multilayer cloak is shown below in simplified form.

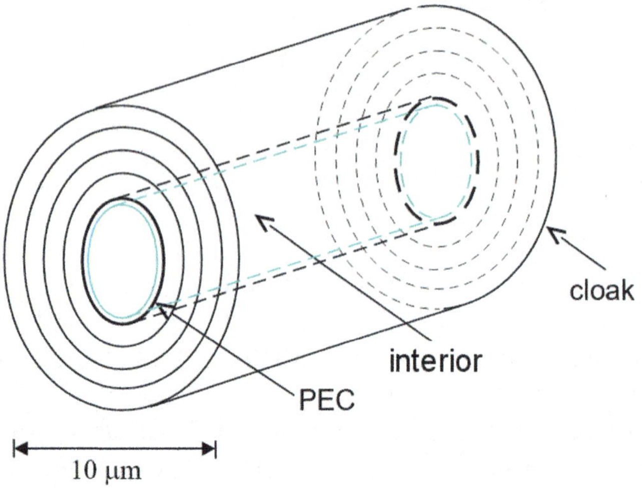

There will be close to *600 million individual parts* (all passive) for the RGB display cube at 128-cubed resolution. However subassemblies can be manufactured separately and pre-tested. These subassemblies will then go on to the final assembly.

2.2. LED Perception

We now switch to LED technology and consider possible technologies for the voxels themselves. The three color frequencies (red, green, and blue) will require different LED layers or levels. The light at each level must be bright enough to be perceived by the viewer. Nanoscale LEDs are probably too small; microscale parts should suffice. The size of an A4 micro-LED is 20 micrometers square. Recall from Chapter 1 that the total width of the light-emitting portion of a voxel area is about 300 micrometers. A 15 times amplification of the A4 LED would fill that area. Of course the dot should be visible to the naked eye, but at the same time each junction of red, green, and blue should be spaced closely enough to create a blended color as perceived by the eye.

The eye is remarkably good at picking up detail so that even a 20 micrometer dot can be discerned. One further concern is that the voxels are space-filling, and don't have too much empty space between adjacent locations. The 2-inch cube size is a tradeoff between the detail of the image and the required space for the millions of transmission lines.

2.3. Micro-LED Technology

The leader of the micro-LED revolution is the Japanese printer manufacturer Oki Data Corporation [6]. They've developed a thin-film bonding technique called Epitaxial Film Bonding (EPI). For 600 dots per inch (dpi) applications, the LED size is A4, with a light-emitting region 20 x 20 micrometers, and an LED array pitch of 42.3 μm. A scanning electron microscope image of this setup is shown below.

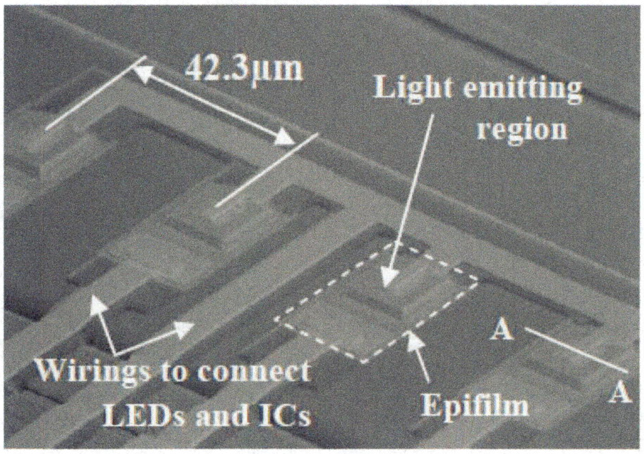

(courtesy [6])

The LEDs are manufactured on one substrate, and this is matched to a CMOS driver that's manufactured separately. The two are mounted side by side and mated with gold wires as shown above. The EPI bonding process works on flexible films, glass, and diamond-like carbon (DLC), this latter acting as an effective heat sink for the LED array. A picture of a 24 x 24 array bonded to flexible film is shown below.

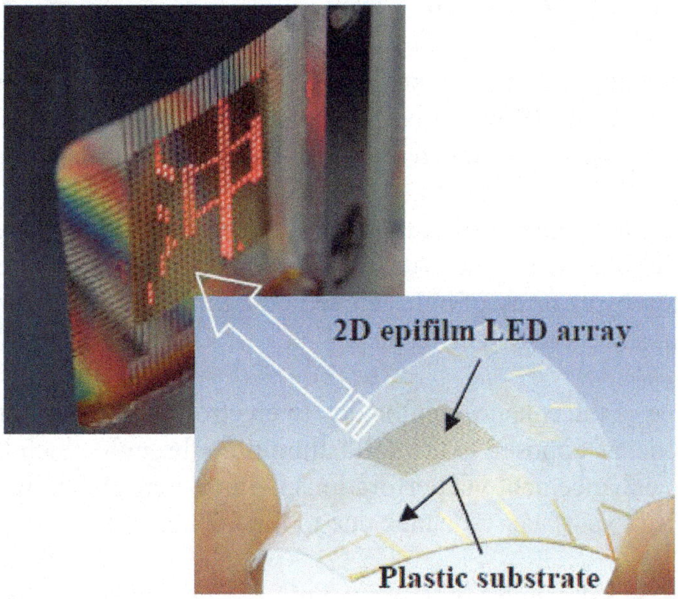

(courtesy [1])

Oki manufactures a 1200 dpi printer that uses even smaller micro-LEDs with a 10 μm x 10 μm light emitting region, e.g. A3 size. These have also been made using the EPI bonding process.

2.4. Micro-LED Technical Details

Lee and Tu have written a comprehensive survey of micro-LED technology [7]. Most of the LED light emission is frustrated by total internal reflection, or TIR. The light extraction efficiency from the surface of the active region of the LED is typically 4.6 percent. TIR is mostly due to the difference in refractive index between the LED surface and air, which causes light rays beyond a certain critical angle to be reflected back into the substrate or film medium. This is shown in the diagram below.

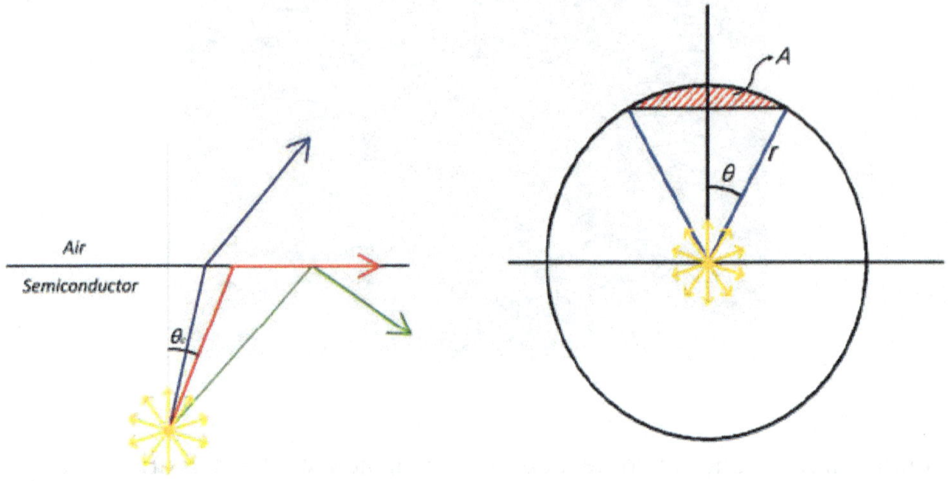

(courtesy [2])

There are different ways to circumvent TIR, but in general the idea is to roughen the surface a little to increase the diffraction and to perturb the angles of the escaping light rays. Lithographic techniques can be used to add bumps that increase light emission.

It should be mentioned at the outset that the cube, though exposed to air on the outer surfaces, will be sealed internally, which means that TIR because of an air-interface won't be the main problem. But there will be other problems, which will be presented in the course of this book.

At the sub-micron level there are some physics that are not very well measured and thus not well understood except by their effects. One such effect is *plasmon resonance*. A paper by Liu et al [8] describes a hole-array technique, whereby an array of holes etched on a thick aluminum film has the effect of increasing LED electroluminescence (EL) efficiency.

The author admits that some of the enhancement is due to electroluminescent transmission through the perforated aluminum cathode, as opposed to the solid aluminum through which there's practically no transmission. But the basic enhancement via perforations is further enhanced by "anomalous" transmissions that are related to surface resonance due to the 2D hole array.

Not all LEDs have the same substrate design. For example, the Oki micro-arrays are based on silicon, not gallium nitride (GaN). Liu's LED is "organic" and composed of a semiconductor polymer called PPV, combined with a metallo-dielectric photonic crystal (MDPC). Amazingly, this LED was made in a homemade glovebox at the University of Utah!

The image below shows the perforated aluminum film in the inset, and the layer structure at the top. The aluminum is labeled as MDPC-Al. The "A" and "G" lines indicate the emission enhancement at the "air" and "glass" interfaces respectively, due to resonances in so-called surface plasmon polaritons (SPPs).

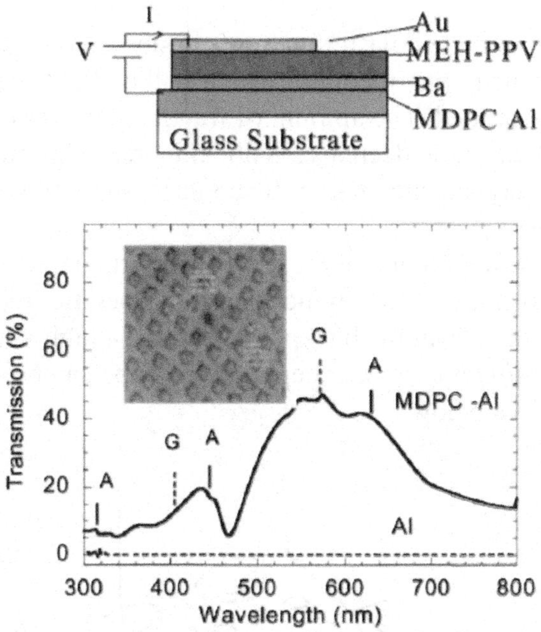

(courtesy [8])

Most blue micro-LEDS are made of gallium nitride grown on a sapphire substrate. However, red and green micro-LEDs have been manufactured using this technology by changing the layer thicknesses [9]. Thus all the colors for an RGB display are available. It should be possible to control the color luminosity via a CMOS driver interface.

2.5. Backplane Waveguide Coupling

The signals that are generated in the cube's cradle, also called a *backplane*, are high-speed optical signals, possibly visible or possibly infrared, which allows high-speed transmissions into the cube's *base plane*, where additional control circuitry resides. When the cube is removed from its cradle, driving signals continue to be generated in the cube's base plane. The base plane is part of the cube itself.

As described in [10], the requirements for a coupling are somewhat rigorous. The main concern is a polished even cut on the receiving surface of the waveguide. Chen and Chen [11] polished the ends of their 200 μm fishing line with sandpaper to enhance coupling efficiency. It must have been very fine 2500 grit sandpaper! Chen and Chen reported waveguide attenuation of less than 0.01/cm.

The polished plastic fiber is mounted in a non-absorbing paper panel over plastic, since Chen and Chen report that paper shows a low absorption of terahertz radiation. To avoid light reflections the paper is matte black. Chen and Chen terminate the light fiber with a parabolic concavity. This trick would be applied on the exterior of the backplane which makes contact with the display cube backface along what is called a *laser strip*. The cube mates to each concavity with a parabolic convexity or *bump*.

Thus the cube's backplane interface looks like a strip studded with low-profile "circular light contacts". This is a risky design since the cube rests in the cradle by gravity, so the pressure will be light between the parabolic surfaces. Any air gap could attenuate the light transfer. But Chen and Chen note that with increased terahertz frequency, the coupling efficiency is improved because of improved *mode-field overlap*.

Chen and Chen are attempting terahertz datastream transmission, and so one must be careful to distinguish between the frequency of the transmitting medium, which will be light or infrared radiation, and the frequency of the data being transmitted. The amount of fractional power (radiation) that flows outside of the polyethylene core of an optical cable decreases with an increase in transmission frequency – that is, with the data frequency. The finely polished parabolic contacts show the same behavior.

The cube, even at a high resolution, will not be receiving terahertz data signals. The frequency of the optical datastream, which is driven by a clock which runs at a specific frequency, is in the low gigahertz range. The cube would receive data from the backplane through a high-speed optical interconnect. This falls under the category of *on-chip photonics*, except that the cube can be removed from the backplane. A sketch of the entire setup is shown below.

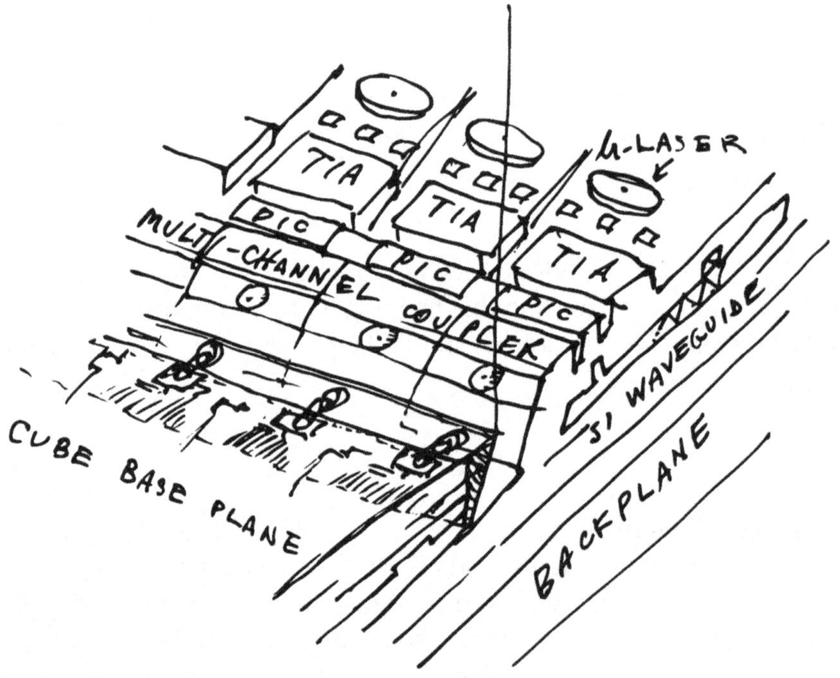

Silicon photonics is an area of ongoing research and it has a language all its own. The coupling between the cube and the backplane isn't trivial. TIA means *transimpedance amplifier*: it converts photocurrent into standard CMOS bits, and presumably goes the other way and converts CMOS bits into a current that drives a laser. A silicon waveguide is etched into a medium that sits on the backplane substrate; it transmits laser impulses to the PIC, or *photonic integrated circuit*. Note that the sketch shows a displaced cross-section of the waveguide: really the waveguide would run directly underneath the laser. The PIC receives impulses from the waveguide and processes the signal with an optical amplifier and optical modulator. The journey from each channel's PIC to the associated photocoupler is short, culminating in concave optical contacts on the 45-degree sloped forward face of the backplane seating area. The cube is shown unseated, pulled back a bit, to reveal the convex mating contacts on the backward sloping face, at the same 45 degree angle as the forward face on the backplane. These contacts connect to *clock recovery stations* in the cube's *base plane* that recover the clock signal of the optical datastream, and thus recover the stream of data.

Typically the type of high-speed circuitry found in the backplane uses *through silicon vias* (TSVs) and a *ball grid array* (BGA) to allow signals to quickly flow from ASIC to PIC. The BGA is designed with a high *coefficient of thermal expansion* (CTE) because optical circuits tend to run hot. In general silicon photonics is a subject of intense research with many challenges.

One final note is that the optical ports must be bidirectional. Each port must receive data as well as send. For example, the data could be corrupt on the cube side, in which case the backplane is told to "resend" the last packet, or in the worst case, to resend the entire *optical plane*. An optical plane is an abstract entity that represents an entire vertically-oriented slice of the display cube.

2.6. Conclusion

This chapter has introduced some major concepts. The cube's data transmission lines have been introduced along with a basic overview of the cloaking technology which hides them. We've taken a first look at micro-LEDs as a possible solution for voxel illumination. As well, silicon photonics has been explored for the purpose of coupling the backplane to the cube.

References

[1] Greenleaf, et al. *Electromagnetic Wormholes and Virtual Magnetic Monopoles from Metamaterials*, Nature, v. 450, November 2007

[2] Hunsperger, R.G. *Integrated Optics, Theory and Technology*, Ch. 2: Optical Waveguide Modes, Springer, 2009, online

[3] Chen, L., Chen, H. *Low-loss subwavelength plastic fiber for terahertz waveguiding*, Optics Letters, v. 31 n. 3, 2006, online

[4] Zhang, J. et al. *Guiding waves through an invisible tunnel*, Optics Express, 2009, online

[5] Zhang, B. and Wu, B. *Cylindrical Cloaking At Oblique Incidence With Optimized Finite Multilayer Parameters*, arXiv:1004.2549v1, 2010, online

[6] Ogihara et al. *1200 dpi Thin Film LED Array by Silicon Photonics Technology*, Electronic Components and Technology Conference, 2008, online

[7] Lee and Tu. *Improving the Light-Emitting Efficiency of GaN LEDs Using Nanoimprint Lithography*, in *Recent Advances in Nanofabrication Techniques and Applications*, Cui (Ed.), 2011, online

[8] Liu et al. *Efficiency enhancement of an organic light-emitting diode with a cathode forming two-dimensional periodic hole array*, Applied Physics Letters, v. 86, 2005, online

[9] Cooke, Mike. *First Yellow-Green and Amber Micro-LED Arrays*, Semiconductor Today, v. 6 no. 10, 2011/12, online

[10] Hunsperger et al. *Parallel end-butt coupling for optical integrated circuits*, Applied Optics, v. 6 no. 4, 1977, online

[11] Chen, L., Chen, H. *Low-loss subwavelength plastic fiber for terahertz waveguiding*, Optics Letters, v. 31 n. 3, 2006, online

Chapter 3

Some Basic Elements

3.0. Introduction

In this chapter some basic elements of the display cube are further developed.

3.1. A Better Way to Blend Colors

An RGB color triple takes the form of a triad of micro-LEDs, one each for red, green, and blue. The 10 micrometer technology from Oki Data Corporation in the form of A3 micro-LEDs is shown below in a 24 x 96 array.

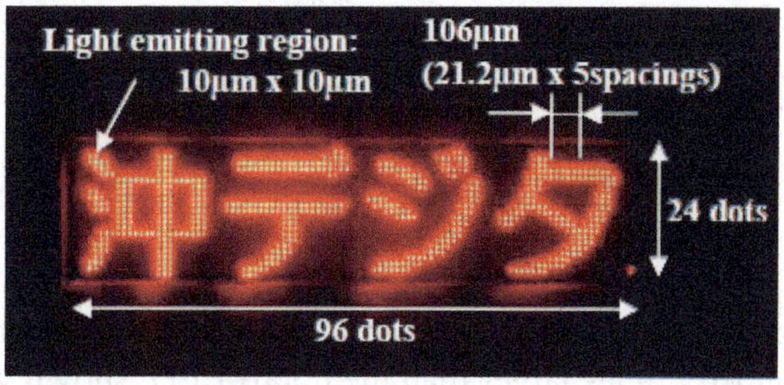

(courtesy [1])

Using the smaller size, the RGB configuration consists of three elements.

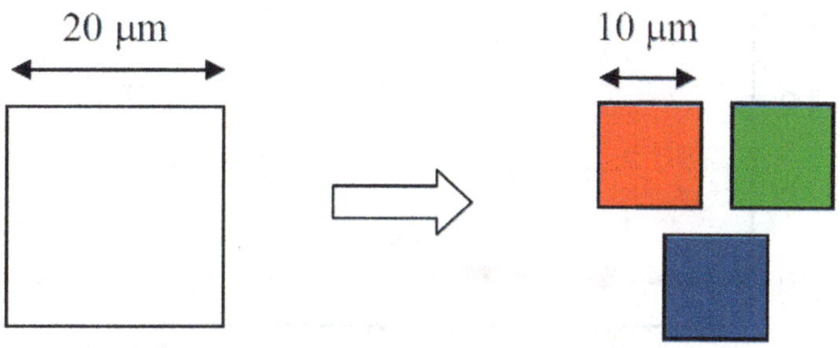

Blending is done by the eye. In the simplest conception, the cube is layered from bottom to top, and the LED clusters reside on each layer.

McKendry *et al* [2, 3] provide the following charts and note that even though the larger pixel sizes provide a larger *absolute* power output, the smaller pixels are capable of producing a higher power output *density*, due to a combination of higher light extraction efficiency, reduced current crowding, and reduced self heating.

The *absolute* power is shown below for various micro-LED sizes at various light frequencies.

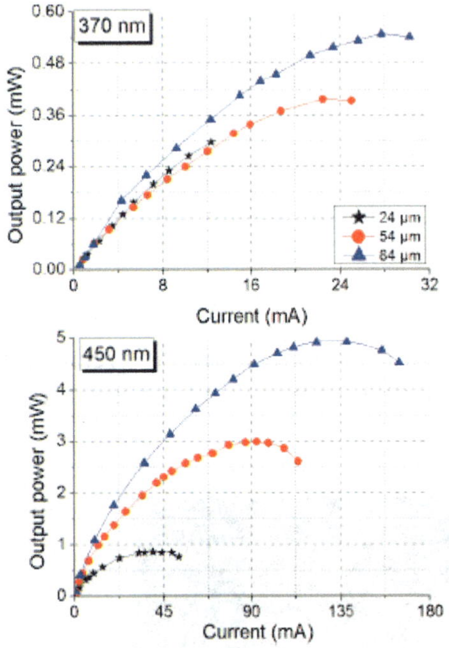

(courtesy [2])

Power *density* is defined in terms of output per unit area. As shown below, the red line represents an LED with a surface area more than fifty times larger than the micro-LED, shown with the black dots. Per unit area, the micro-LED shines brighter.

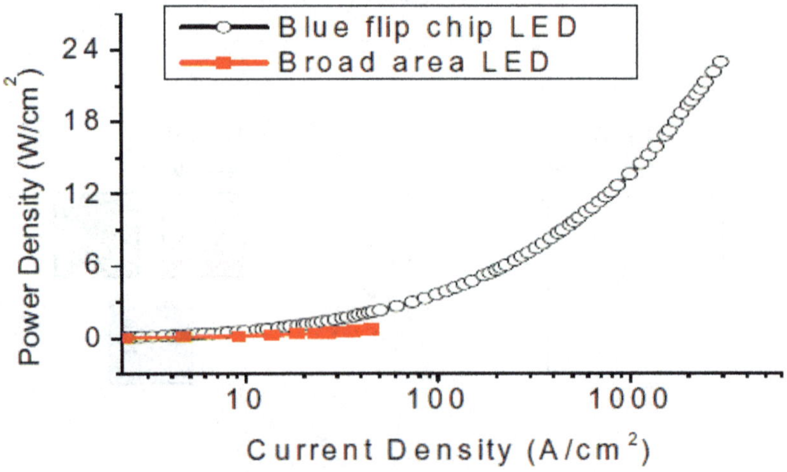

(courtesy [3])

For color blending, each red, green and blue micro-LED will contribute more light per unit area to the final blend. Larger LEDs have a larger area and so a larger *absolute* power output, but if one were to cut a small footprint out of the larger LED terrain, it would perform more weakly than the same-sized micro-LED.

3.2. Layer Management and Manufacturing

The transmission lines travel vertically through the cube from bottom to top. This creates some challenges in terms of the manufacturing process. These challenges will be addressed in the following chapters. The following diagram shows how the cube can be built up as a series of thin layers.

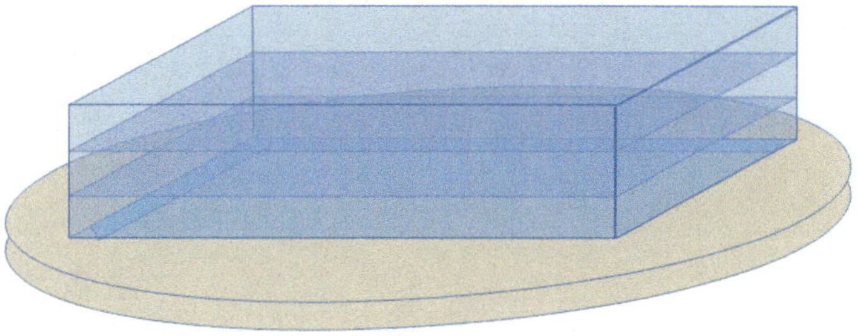

The cube is assembled something like a building, layer by layer. Each layer can be manufactured like a chip wafer. Some layers will have circuits etched with ion-beam lithography. On some layers a substrate of silicon oxide makes sense, on others the medium would be boron glass, or borosilicate. The layers would be fused with optical clarity except where the circuitry is exposed. The layer fusion process would be seamless and ensure that when looking into the completed cube no traces of the layers are visible.

The vertical lines are not extended wires, but stacked contacts that make vertical contact over very short distances, on the order of tens of microns. Therefore the only limiting factor for the vertical conduit density, since it's all etched from level to level, is transparency. The cube's height of 128 voxels is divided by 4 and those 4 groups are written to in parallel. If there are 8 lines per group, then the hope is that 8 x 4 = 32 lines will meet the transparency objective. Given this, the vertical conduit would look something like the following in plan view.

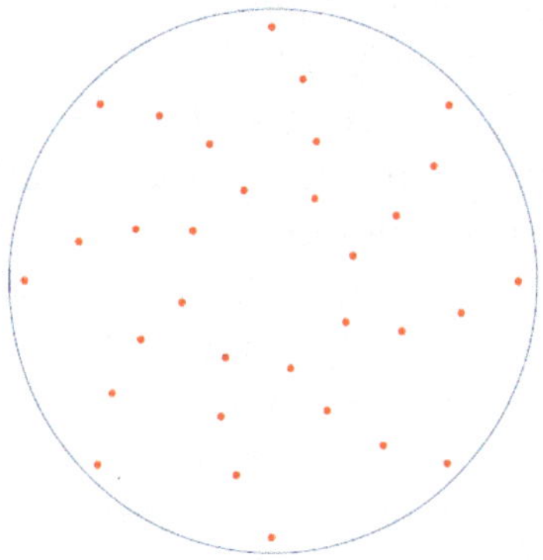

Now, the conduit is very thin, never mind the extremely small scale of the actual lines. But the trick is that each group of 8 stops after its 32 levels are done, so that the conduit has 3 transition levels that look something like the following.

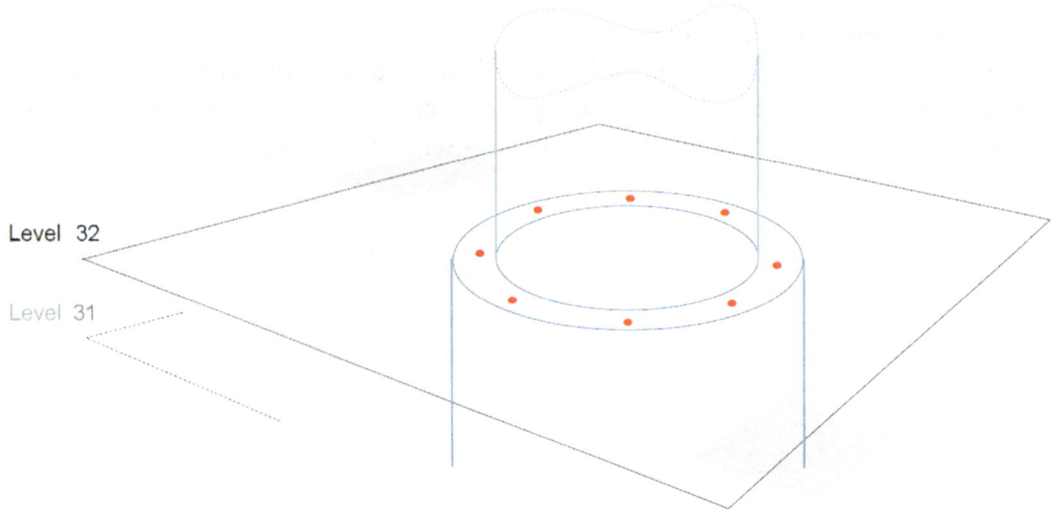

The column becomes more slender at each transition level, so that towards to the top, which is the side of the cube most people will be looking into, the transparency will be greatest because the diameter of the column will be smallest.

3.3. Backplane Memory

The biggest challenge for the design of a solid-state 3D display device, both for the cube and for the backplane, is having enough room to lay out the various components. The backplane must have enough room for the fiber-optic strip and backplane components.

The memory for the 3D display device would consist of a cache used to hold the overall voxel image, contained in the backplane; and a holding register for each CMOS micro-LED driver circuit, contained in the cube at each voxel.

In the simplest scenario each LED is either on or off: each register would simply be a flip-flop that holds on/off state. But for color levels over a range, the micro-LEDs must handle variable resistance, i.e. luminance levels, in which case something more sophisticated at the CMOS level would be needed, such as 3 bits per red, green, and blue. The final design will use the industry-standard 8 bits per RGB. However, the more bits per pixel, the more complicated the control circuitry is.

For the simple 3-bit scenario, given a row size of 128 and a column size of 128, memory is laid out in 32-bit increments. Each 32-bit word resides on a single memory chip. Four chips take care of the *columns*. Each row would be read out in "rapid fire," 128 rows in all. Memory is arranged in *row* and then *level* order. Each level is offset by 128 rows in memory, i.e., the next 128 rows for level 2 would be contiguous in memory; and then the next 128 rows after that; and so on. This is shown schematically below.

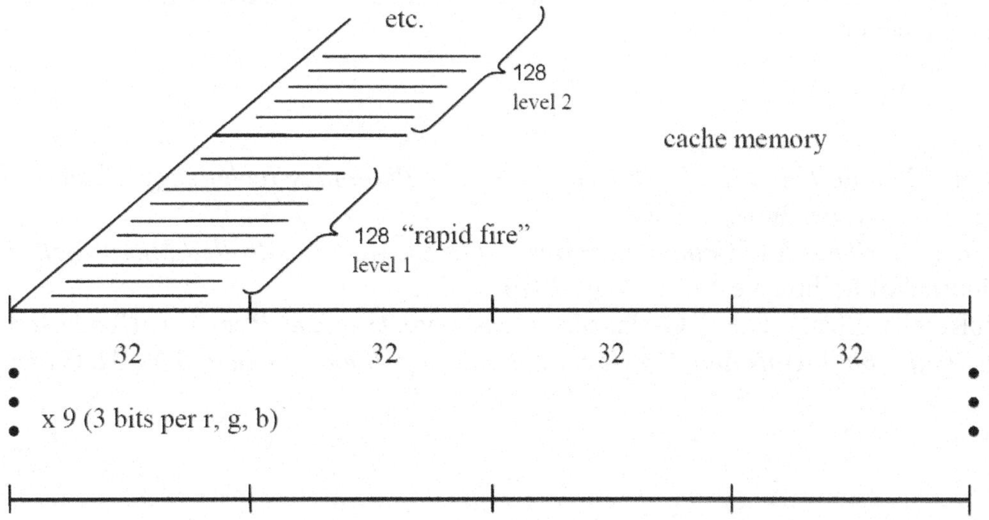

Notice that there are 9 x 4 = 36 chips, which can be laid out in three 3 x 4 arrays. Each chip only needs to hold 128 rows times 128 levels, or 16,384 bits. This is further multiplied by 32 bits per chip, and divided by 4 bytes per 32 bits, to give 131,072 bytes or 128 Kbytes per chip. A total of 36 chips adds up to 128 x 36 = 4.608 Mbytes total memory usage. Note that for a 1-bit (on/off) implementation, only 4 chips as above are needed. This is the simplest kind of storage setup. However, the actual data format will be more complicated because of the conversion to a serial stream and the use of packets. (More on this in later chapters.)

At this early design stage it's not necessary to specify a hard mapping of memory lines to optical ports, since the number of optical ports, as well as the width of the color data may change as the design develops. Whether 9 or 24 color bits are available, or whether the column addressing and row addressing should be swapped, can all be locked down later.

The refresh bandwidth for 128 rows is 128 x 60 Hz = 7680 Hz. To clock out all the levels (and so return to the first level at a rate of 60 Hz) requires an additional increment of 128, giving 7680 x 128 = 983 kHz. This is the gate-out speed for the memory chips, but it turns out that the speed will be much faster because the data will be transferred to the display cube in *bursts*.

3.4. Conclusion

This chapter has outlined more of the design for an RGB 3D display device. Features of the developing design are:

- Panoramic visibility: five out of six sides of the cube visible.
- Color blending on each layer of the cube.
- Use of horizontal layers.

However, there are still many challenges. Where and how the display data is stored in the cube will be an ongoing topic of discussion. The data can be stored in registers in the base plane, or there can be holding registers at each nexus of LEDs around a central transmission column on each level. In fact, both approaches are valid. The latches won't hold all values for all LEDs at once. These concerns will be addressed in later chapters.

References

[1] Ogihara et al. *1200 dpi Thin Film LED Array by Silicon Photonics Technology*, Electronic Components and Technology Conference, 2008, online

[2] McKendry et al. *Visible-Light Communications Using a CMOS-Controlled Micro-Light-Emitting-Diode Array*, IEEE Journal of Lightwave Technology, 2010, online

[3] Griffin, McKendry, Zhang, Gong, Guilhabert, Massoubre, Gu, Rae, Renshaw, Henderson, Dawson. *Micro-pixellated flip-chip InGaN and AlInGaN light-emitting diodes bonded to a CMOS driver backplane*, 2011 (poster)

Chapter 4
Packet Switching

4.0. Introduction
This chapter delves into data bandwidth.

4.1. Laser Ports
The design as it has been presented up to now consists of a backplane that contains 128 x 128 = 16,384 RGB triples, one triple for each voxel in the plane. To display colors in the z-direction (vertically) the backplane cache cycles through the color data memory for each vertical column.

The optical port can be driven by visible light or radiation in the infrared range. Either way, the radiation needs a higher Q or *quality* than is available from scattered LED light. Usually coherent laser light is used. The idea is make all of the fiber-optic sources micro-lasers.

4.2. Data Modulation
To get data to the LEDs the coupling between the backplane and display cube must transmit actual data. The "rapid fire" memory layout as covered in the previous chapter still holds, but the new sense of things is as follows.

Each pin on the cache pinout can be viewed as one bit of a serial stream of data, that, for that one bit, proceeds for each memory location in *column, row* and then *level* order. Thus 128 rows for 128 columns are strobed out for level 0; and then the next 128 rows for 128 columns for level 1; and so on. This ordering assumes that the full cube is being refreshed.

Data is serialized and modulated onto the fiber-optic strand, but how is it encoded? How does the data reach the right column in the display cube?

4.3. The Manhattan Scheme
One model for data packet routing is called the Manhattan Street network. Each node has two lines *in* and two lines *out*, from and to neighboring nodes. The topology wraps around on itself from column *top* to column *bottom*, and row *left* to row *right*, making the topology toroidal. The various routing algorithms that use this network take advantage of the various links to route arriving packets in the least amount of time possible.

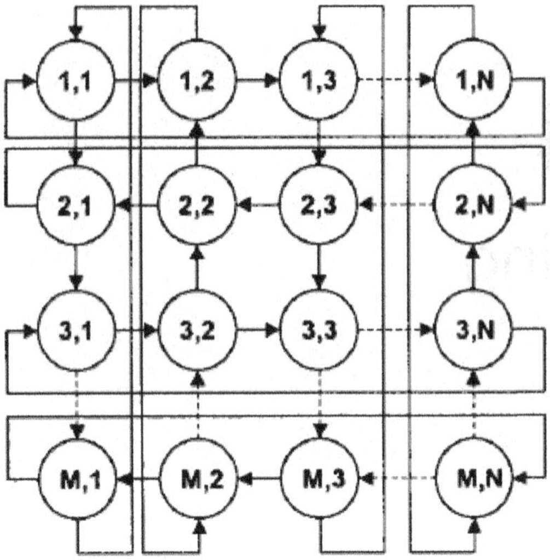

(courtesy [3])

The nice thing about this routing network is that it's two-dimensional, much like each plane of the display cube. The addressing takes the form of a *row* address and a *column* address. The encoding scheme is shown below.

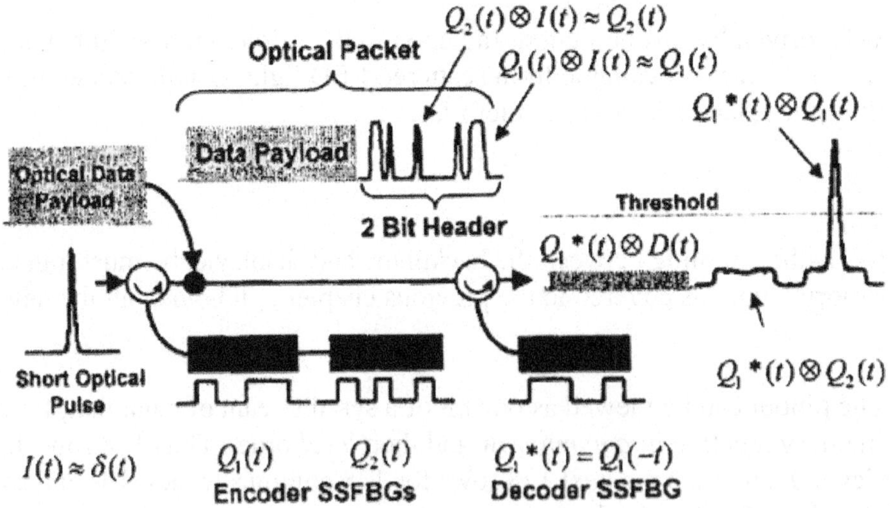

(courtesy [3])

The approach uses Superstructured Fibre Bragg Gratings (SSFBGs) [3, 4]. Each receiving column relies on *optical correlation* to compare each bit of the data with the address encoded by optical reflection onto an encoder grating that's been pre-etched with a unique spatial refractive index profile. The optical label is said to be *phase encoded.* Each bit is then decoded with a second Bragg grating, identical to the first, except illuminated from the opposite end to create a time reversed impulse response, which creates a peak in the received signal, thus signifying a match on the bits. The flipped Bragg grating is signified as Q* above, which is equivalent to Q(– t) rather than Q(t). In essence it's the reverse principle but abstracted into the refractive index domain.

4.4. Applying the Bragg Scheme to the Display Cube

Packet switching schemes are designed for long distances. Packet switching papers typically have statements like "the packet stream was transmitted over a distance of 25 km of standard single mode fibre." [4] Even though packets travel enormous distances, the switching concepts apply to instantaneous parallel processing of node addresses. The display cube is a miniature grid, and so Bragg gratings can be used at the nanoscale to essentially do all the work. But manufacturing gratings is expensive, and the sheer number of address decodings in the cube would prohibit their use. However, optical encoders may only be needed in the cube's base plane, where circuitry guides signals to the correct column.

In the packet switching world a data packet is *added to* or *dropped from* the network relative to a local node. A packet that arrives with the node's address is dropped from the network; the data goes directly to the node to be processed as required. The node can also add a data packet to the network. The display cube will mostly receive packets, and send only when requesting a new packet stream. As well, the cube will *forward* data from one column to the next.

In the Manhattan scheme there are two inputs from neighboring nodes, and two outputs. The simplicity of the display cube suggests that the Manhattan routing can be simplified quite a bit for 3D display purposes.

In essence the display cube becomes an elaborate optical memory. It's not necessary to scan out voxels in column, row, and then level order, although for a refresh update this is the preferred order. This would be called a *full update*. An *incremental update* based on a specific row and column address for some plane is an alternate way of addressing memory. However, this would mean that the other locations continue to be refreshed independently – and that would complicate the cube's routing and address decoding as well as its memory footprint.

In general, instead of two inputs for each node, as in the Manhattan Scheme, there's simply an address decoder. Data forwarding is limited to a *column header / controller* node; at destination voxels there's simply either a drop, or no action.

The data packets are timed to a clock. During each time slot a node can accept a packet. The routing architecture is based on the *column* then *level* memory layout of the display cache. As in the Manhattan addressing scheme, a *row address* and *column address* precede the data.

A two-stage simplification of the Manhattan scheme is as follows. The first stage is the column address decode. Each packet simultaneously reaches every column header / controller node, 128 in all connected to each cache output pin on the backplane coupling strip. The column address is decoded at the controller node.

At the same time, all of the row elements extending vertically will also get the data packet on the clock pulse. However none of them will match on a row address as initially encoded.

At this point we have to decide what gets converted to an electrical signal and what remains optical. If everything remains optical then a packet could be processed as follows: each address is preceded with a *start bit* used for synchronization. A match at the column header results in the following actions: a) the column address is dropped; b) the header is split into two signals; c) the start bit is duplicated, delayed and appended to the original start bit; d) the row address is appended to the two start bits; e) the new address packet is forwarded up the column.

The second stage is completed when a row element matches the forwarded address header. Each row element matches on *two* start bits. The latency for a received packet is one clock cycle, i.e. a single wait period for the column address decode and row address forward by the column header / controller.

The problem with this approach is that high-speed signals are leaving the base plane and being sent up each column, making the whole cube, in essence, an optical data network. This is not desirable for several reasons. Instead, the high-speed optical signals should be limited to the cube's base plane and converted at the base plane into electrical signals that then are sent up the columns to the color planes.

In the simplest design, each column of the cache memory is modulated by a laser into an optical serial data stream. The encoding relies on bit position and bit value. The addresses are synchronized with a start bit. On the cube side, bit 0 goes to Bragg input 0, and is split off either into a Bragg <0:0> or <0:1> encoding. 128 columns are encoded with 7 bits, and so there are 7 x 2 = 14 Bragg encoders per grating. This makes 14 + 1 start bit = 15 Bragg encoders for each column. This means there are 128 x 15 = 1920 Bragg encoders total per *optical plane*. There are 128 optical planes, and so 1920 x 128 = 245,760 encoders!

The Bragg technology is simply etched fiber. Since the encoding is shallow (14 bits) the fiber doesn't need very many *chips* – the name for the etched sub-pulse gratings on the fiber. The bigger expense is splitters and delays. Manufacturing the base plane would also be complexified by the use of Bragg gratings.

A conceptual diagram for the data paths is shown below.

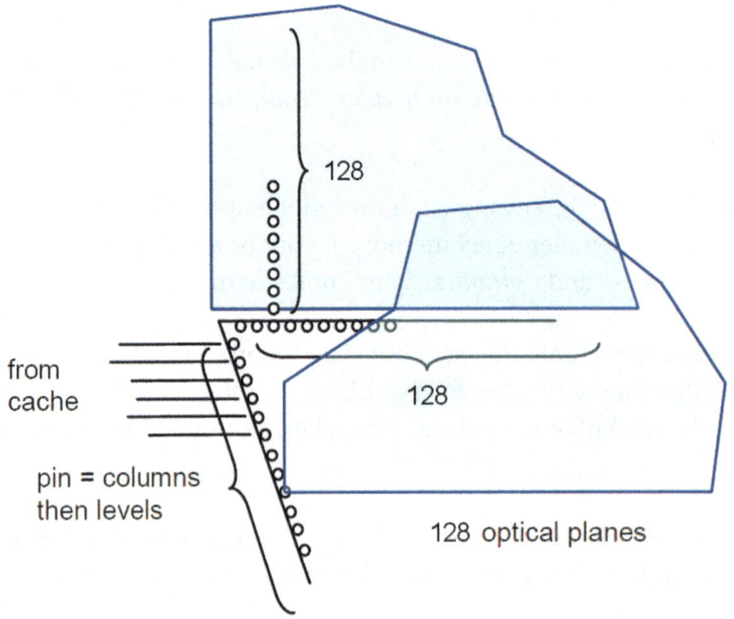

An *optical plane* is oriented vertically, but it's an abstraction of convenience, to give a visual representation of the backplane memory layout. Let's continue with the optical design even though it's expensive. Note for now that the alternative is to simply take advantage of the sequential nature, from column to column, of the optical plane associated with each optical coupler, and use cascade signals.

4.5. Bragg Scheme Continued

The circuitry for each column controller requires some delicate timing. The address header is "picked off" in an optical coupler for address decoding; meanwhile the data payload is sent directly to an electro-optical switch after being put through an optical delay to compensate for the address decoding processing time. The decoding is done in parallel, as shown below. A group of "on" pulses is generated for a perfect match.

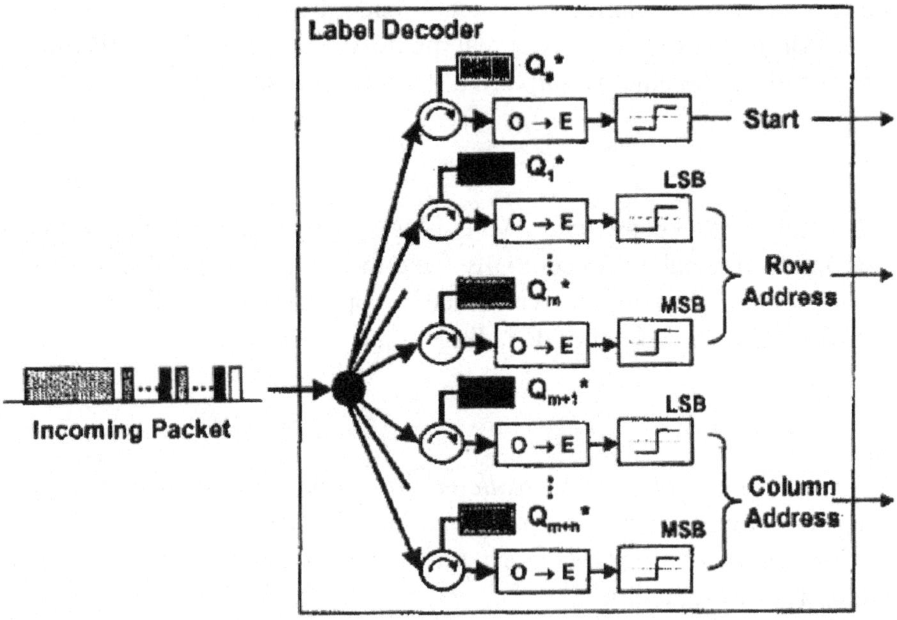

(courtesy [3])

Parallel decoding amounts to feeding the serial packet into separate bit decoders, each one with the reversed Bragg grating corresponding to the Bragg encoder grating. The "O -> E" notation stands for "optical to electronic". Parallel decoding is useful therefore to extract the bits from the serial stream, and would include a conversion to some kind of electrical signal. In the fully optical interconnect design, the column headers would each have a grating for the column address; and as well a grating for a row address. But conceptually there are no rows in the base plane. The cache pinout is in fact a physical row connection, so no row addressing is needed. We'll see in Chapter 6 that *level* address decoding doesn't require optical encoding either.

Each column header is completely ignorant of its address since the address is encoded in the Bragg gratings. Each address bit, once decoded, will either generate a peak (a match) or no peak (no match). The peak is called an *auto-correlation peak* (ACP). An address match is determined with a logical "and" on all of the decoded bits. A logical output of "1" enables an electro-optic switch at the column header to switch through the data packet to the column controller. Otherwise nothing happens.

What happens once the data is switched through? This will be the subject of Chapter 6.

In [3] the Bragg encoding is fed with optical pulses of about 20 ps each, reflected off the individual Bragg gratings, producing encodings per bit on the order of a few nanoseconds. The optical pulses are driven by a CW (continuous wave) laser.

For the display cube the modulated pulses will be pumped by a microlaser at each of the 128 cache output pins, e.g. the optical couplers. One might think that pumping is unnecessary on a toy like the display cube, with spans of millimeters rather than miles. But we will see in the next chapter that nano distances have reduced power ratios, and though the pumping is weaker, it's still necessary. The laser power has to be strong enough to carry a signal that fans out to 128 column controllers; from there, a digital signal is guided vertically up each column.

The data can be encoded in any format since it's just a bit code, but for high-speed transfers a spectrally efficient format such as non-return to zero (NRZ) may be best. Note that all 9 RGB bits are available at the display cache in parallel and can be quickly converted to optical pulses.

4.6. Conclusion

Here we have looked closely at packet switching technology in the context of feeding RGB data to the individual columns of the display cube. Conceptually the cube contains no rows, but it's wise not to lock down the abandonment of row addressing just yet, since the optical planes and thus the rows may need to be *demultiplexed* and so require some kind of decoding.

References

[1] Koppa et al. *Demonstration of optically controlled data routing with the use of multiple-quantum-well bistable and electro-optical devices*, Applied Optics, v. 36 no. 23, 1997, online
[2] Bagnoud and Collings. *Study of a bistable GaAs Fabry-Perot device for operation in array format*, Optics Communications 114, 1995, online
[3] Thomsen et al. *Self-Routing in a Photonic Packet Switched Manhattan Street Network based on all-optical label recognition*, Photonics in Switching Conference, 2003
[4] Teh et al. *Multi-wavelength (40 WDM x 10 Gbit/s) optical packet router based on superstructure fibre Bragg gratings*, IEICE Transactions / Joint Special Issue on Recent Progress in Optoelectronics and Communication, 2002, online

Chapter 5

Timing Considerations

5.0. Introduction

Here we delve into timing issues.

5.1. Serial Bandwidth

As outlined in previous chapters, memory is laid out in column then level order. Moreover, each bit-output of display cache memory corresponds to an *optical plane*. There are 128 bit-outputs, all read out in parallel.

Thus, each optical plane is processed in parallel with a bandwidth of 128 x 128 x 60 Hz = 983,040 bit reads per second. However the optical plane is updated with a stream, with 3 bits per red, green, and blue encoded serially. The total bit rate is thus 983,040 x 9 bits = 8.847 Mbits per second. It will be a little higher when taking into account the encoding of the column address that precedes each data packet.

Telecommunications bandwidths are in the Gbit per second range. Higher display resolutions may approach this higher bandwidth. For example, a 512-cubed display device has an optical-plane update rate of 141.55 Mbits per second. Therefore it would seem that optical telecommunications bandwidths are well-suited to larger voxel resolutions. Note that the design only involves an *n*-squared increase in bandwidth, not *n*-cubed.

5.2. Non-Return to Zero Format

A bit rate of 10 Gbits per second translates into a pulse width of 100 picoseconds, as shown below.

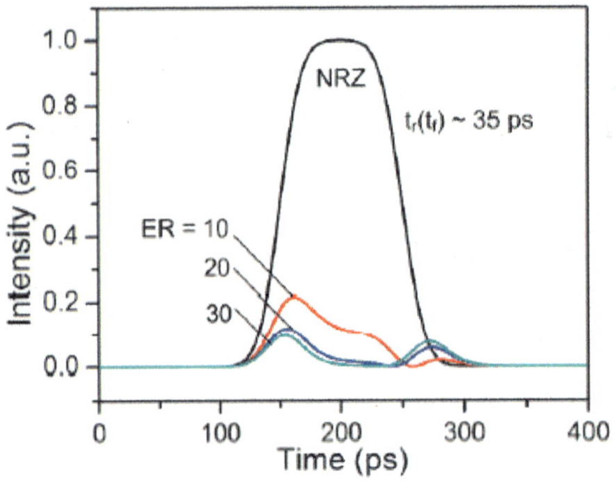

(courtesy [1])

The authors of [1] modeled each pulse with a rise time of 35 picoseconds, and the same for the fall time. According to [2] the NRZ encoding decomposes in the following way.

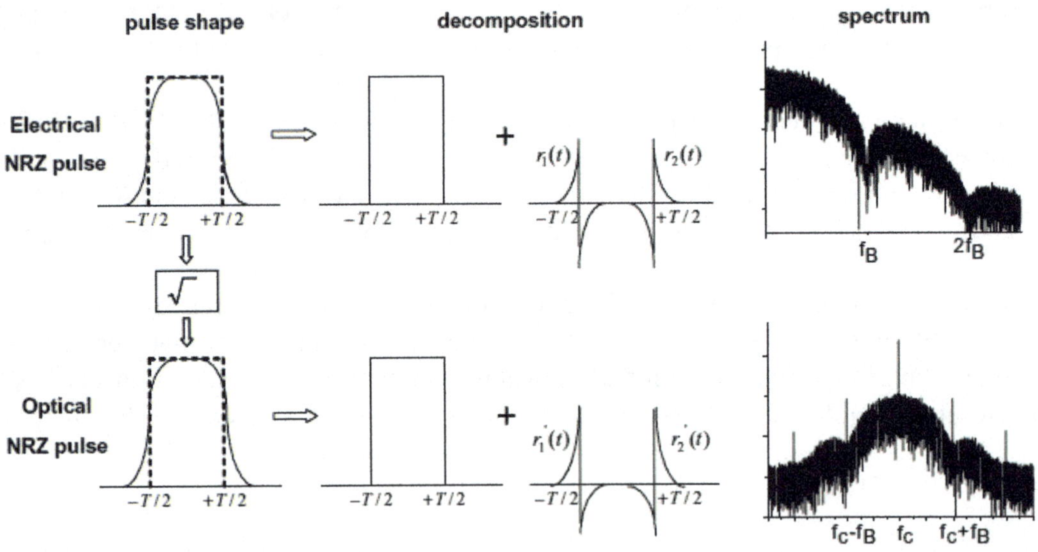

(courtesy [2])

T is the inverse of the bit rate, or equivalently the clock frequency. A bit rate of 10 Gb / sec, or a clock frequency of 10 GHz, has a time T of 100 picosends. Thus the NRZ pulse as shown in the first graphic extends from -50 ps to 50 ps. This is the *ideal* square wave NRZ pulse. As well, there is a finite rise and fall time shown as r(t), or t(r) and t(f) in the first graphic. According to [1] this adds 35 ps x 2 = 70 ps to 100 ps to make a total pulse width of 170 ps.

The non-return to zero or NRZ format is encoded as shown below.

Usually the 1's are encoded with a positive voltage, and the zeroes with a negative voltage. This is called a *bipolar* encoding. The encoding never has a neutral or zero state, which is why the encoding is "non-return to zero." As well, there's no clock encoding in the NRZ format. Due to the finite rise and fall times, optical clock components can be extracted at integer multiples of the clock frequency from the carrier center frequency.

The digital data is intensity modulated into the pump laser stream which acts as the carrier signal and operates at a fixed frequency. In [1] a carrier wavelength λ_c of 1550 nm is used. Buried clock signals tied to the NRZ bit rate are offset from this center frequency in steps of 10 GHz, as shown below.

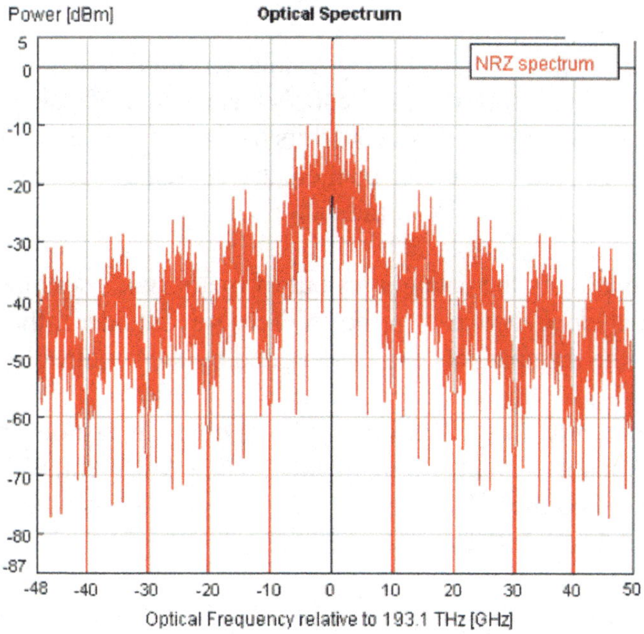

(courtesy [3])

Note that 193.1 THz can be inverted and multiplied by the speed of light to give the carrier wavelength of 1550 nm for the pump laser. The [GHz] in brackets refers to the horizontal scale at integer multiples of the bit rate of 10 GHz. These vertical divisions are called *tones*. This spectrum is also shown in the decomposition graphic above, with the tones shown as offsets from the carrier frequency f_c.

Note that clock recovery isn't possible in the electrical spectrum because the rise and fall times are symmetric enough to completely cancel each other out. But when the *electrical* NRZ signal $S_e(t)$ is fed into an optical intensity modulator to modulate a continuous wave (CW) laser, the intensity of the modulated signal is proportional to $S_e(t)$, while the amplitude of the modulated *optical* signal $S_o(t)$ is proportional to the *square root* of $S_e(t)$.

This nonlinear response destroys the symmetry of the rise and fall times and the residual pulses remain intact in the optical carrier. This then opens the door to recovering the clock from the optical NRZ signal at integer multiples of the bit rate.

5.3. Simplification to Super-Structured Fibre Bragg Gratings

As outlined in Chapter 4, the optical transfer scheme involves two address decodings using Super-Structured Fibre Bragg Gratings (SSFBGs). In the naive approach over a million SSFBGs must reside in the cube to decode every voxel location.

However this design can be simplified by extending the data packet to include all of the data for every single level for some column of an optical plane. Then, each level decodes the data not with a row address header that's forwarded by the column header / controller, but by simply reading a data bit at the proper bit offset in the data packet.

In this way the address headers are only generated for the column headers / controllers. This means that SSFBGs are only used at the base plane level by the column headers. In turn, this reduces the SSFBG count to 128 x 128 = 16,384 x (7 bits = 128 values for 128 columns) = 114,688.

However even this simplified scheme has one failing: miniaturization. The Bragg gratings won't scale down easily to fit the 2-inch square footprint of the cube's base plane. We take a cue from Chapter 4 and realize that the Bragg encoders can be used to decode bits as part of the demodulation process, so that the serial stream is both deserialized, and converted to an electrical signal – not at the column headers, but at the coupling stations in the cube, farther back in the data path.

More improvements can be made. As the data comes out of memory it's serialized to a high-speed data clock, and the main concern is *clock recovery* on the receiver side, inside of the cube on the base plane. After the clock is recovered, the data packets can be decoded by the "control logic" in the base plane.

High speed data lines are inherently noisy. A differential scheme reduces noise by balancing the transistor currents with two inputs: the positive input and its inverse, e.g. a "differential". The landing pad for the incoming serial stream, situated on the cube side of the optical coupler, is where the clock signal is recovered, as well as the digital serial stream.

In Chapter 12 we'll see how the vertical conduits can be placed at every other voxel and an optical strip with 128 optical couplers can be cut in half. The number of column headers can also be cut in half. This means that each "station" handles a *quad*: a footprint of 4 voxels. However a further improvement can be made by halving the stations yet again, to handle 4 quads, or 16 voxels. This reduces the number of optical ports to 32. Each port takes the serial data and converts it into four separate signals, by using a 4-phase multiplying delay-locked loop.

The signals are labeled for the processing units residing in the color planes. There would be four processing units: left top (LT), left bottom (LB), right top (RT), and right bottom (RB). At each of the 32 optical ports, the four signals are carried by differential link lines to alternating centers, left and right, each center corresponding to a double quad of voxels distributed between four optical planes – the abstract planes that are vertical slices of the cube.

Most of the link lines are flowing in the same direction which makes the link unidirectional. However, a single Ack/Nack line communicates with the landing pad at the optical port, which makes this design a point-to-point bidirectional design.

What's just been given is a preview of the final design. Let's return to the less complicated design for simplicity.

5.4. Synchronous NRZ Format

Say that the column address takes up 7 bits, preceded by a synchron or sync bit, making 8 bits. There are 128 column addresses x 60 Hz = 7680 x 8 bits (address header) = 61,440 bits per second just for the column headers. The real bandwidth comes from the data. For each column the data packet would contain 128 x 9 bits = 1152 bits. For a refresh rate of 60 Hz we have 1152 bits x 60 Hz = 69,120 bits / sec. This in turn happens for every column, and thus the total data bandwidth for the optical plane is 69,120 x 128 = 8.847 Mbits per second.

Adding in the column addresses makes 8.8908 Mb / sec. However, there will be some spacers between data packets, so to round things off the bandwidth can be set to 10 Mb / sec. This then is inverted to get a NRZ pulse width of 100 nanoseconds. This then is 1000 times slower than the telecommunications bandwidth given in [1] of 10 Gb / sec. Though slower, this longer pulse width can still be sent over a carrier of 1550 nm, as above.

5.5. Extracting the Clock

As outlined in [1], two micro-ring resonators can be used to extract an NRZ clock signal from the carrier. The setup is shown below.

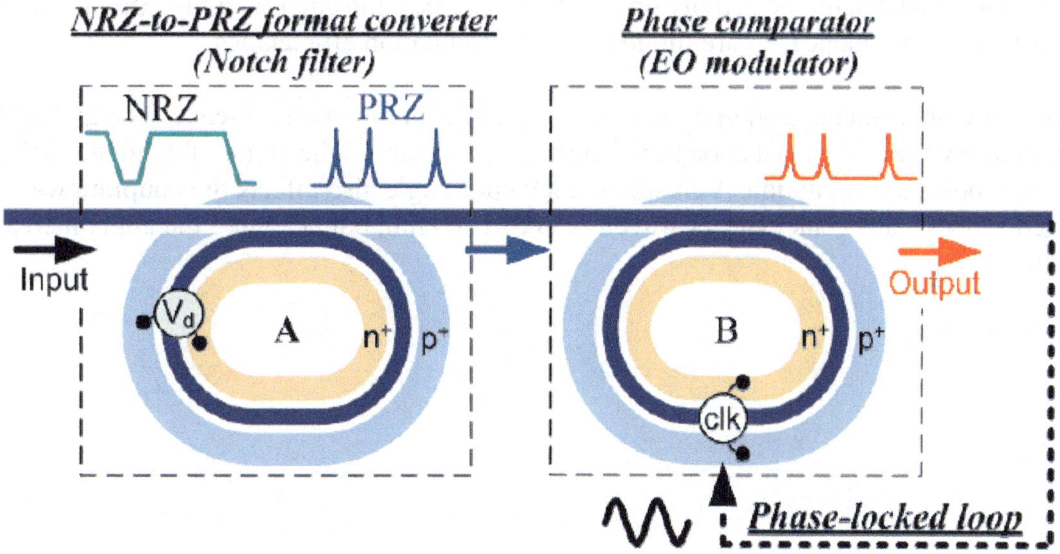

(courtesy [1])

A *pseudo-return to zero* or PRZ signal is simply a pulse aligned with each NRZ transition. The PRZ is then run through an electro-optic (EO) modulator, e.g. micro-ring resonator B. Resonator A converts from NRZ to PRZ and is driven with a fixed voltage. Resonator B is driven with a variable clock signal which acts as a periodic voltage. The phase of the clock signal causes the intensity of the output to be attenuated or amplified, depending on how close the clock signal is to the PRZ, which acts as a reference input. Thus resonator B acts as a phase comparator; meanwhile, phase detection, i.e. detecting the difference in signals, is done in a CMOS micro-chip. The clock can be sped up or slowed down when the phase begins to drift, which is a measure of how attenuated the output of resonator B becomes.

This phase-locked loop (PLL) can be run for the entire optical plane, for all column headers, at a tiny *clock recovery station* (CRS) on the base plane of the cube. The CRS serves all 128 column headers of the optical plane with a clock signal. This station would include the CMOS phase-locked loop chip that modulates the clock frequency; and the dual resonator setup above. For 128 cache pin-outs corresponding to 128 optical planes, there would be one clock recovery station per pin, making 128 clock recovery circuits and 128 clock feeds, one per *optical plane*.

Once a clock is available it can feed the entire plane. As described in Chapter 4, a column address is decoded and matched at a single column header / controller; and this causes the data to be forwarded up the column. The data is forwarded for all 128 levels.

In short, a sync bit at the data packet head starts a counter at the column header / controller (7 bits for 128 counts). The count corresponds to each level's bit position in the data stream. The counter output is fed to a digital multiplexer. Each level is activated in turn to gate through an NRZ data bit. The timers loop through to the next bit, until all 9 bits have been decoded for all 128 levels.

5.6. Carrier Configuration and Incident Geometry

Caspers *et al* [4] have devised a grating coupler that takes advantage of plasmonic resonance to boost the light signal. This grating or one like it would pick up the modulated beam on the cube side. A modulator based on a microring resonator can suffer from photon interference because the resonator Q value is too high. Caspers notes that plasmonic activity on thin films can be controlled with high precision, but that they require either high pump fluences or are limited by low modulation efficiencies.

Caspers came up with a silicon-gold grating to couple light with low pump fluence. The grating has a picosecond recovery time, so is fast enough for Gbit/sec processing. The angle of incidence affects the surface plasmon polariton excitation. A change in incidence angle also affects the coupling wavelength. Caspers tested a device at a center carrier wavelength of 1400 nm, but changing the angle increased the wavelength to 1550 nm.

The setup is shown below.

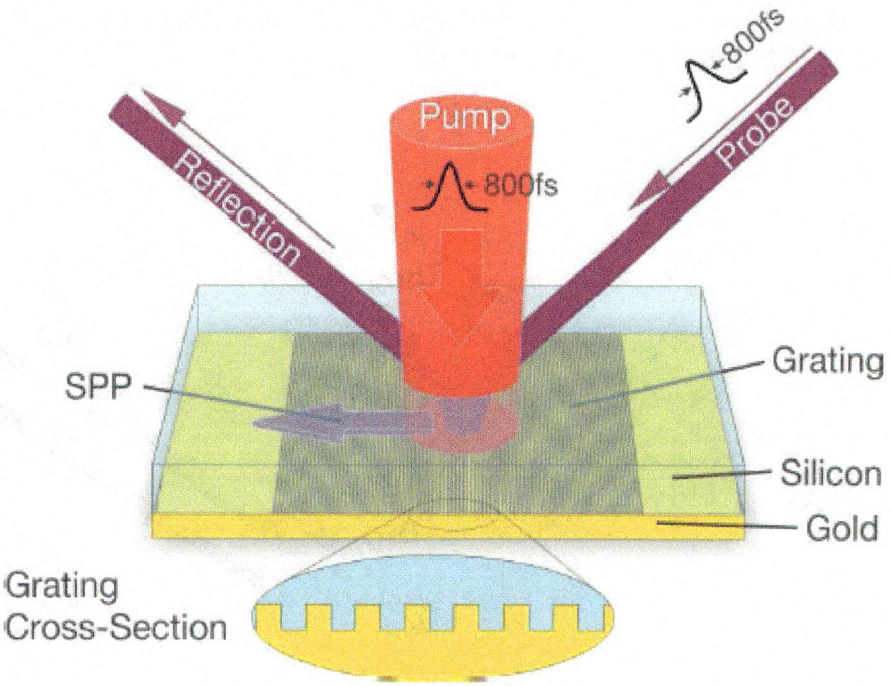

(courtesy [4])

This setup is experimental and so lacks integration. Ideally a microring pump laser would be integrated with a waveguide, along with a microring modulator as hinted at in Chapter 2. The coupling grating has a period of 500 nm and its overall dimensions are 500 μm square. This seems large, but it nonetheless fits within a single voxel span, assuming that these couplers are situated along the *laser strip* (also called a coupling strip) that connects the backplane with one of the cube's lower edges. Note that the pump lasers would reside in the backplane, not the cube, and would couple to the cube along the laser strip for all 128 pin locations.

The spot size of the probe laser is 100 μm in diameter. The probe signal is driven with an ultrafast titanium-sapphire laser impinging on an optical parametric amplifier (OPA). The center wavelengths of the OPA/ pulse laser setup are between 1300 and 1700 nm, which corresponds to 800 femto-second probe pulses, producing surface plasmon polariton (SPP) resonances at 1400 and 1550 nm, in the near and mid infrared frequencies.

The pump laser, which is the optical carrier, has a center wavelength of 775 nm. This laser pumps a 0.5 μm layer of silicon, and has an absorption depth of about 9 μm. The pump laser spot, aimed square at the grating, is 500 μm in diameter, and so covers the entire grid. The energy density for the pump laser is given as 2.2 mJ / cm^2, and after a lengthy calculation this works out to a power flux of about 63.7 mW over the spot area.

Typically microring lasers generate power on the order of about 300 nW. However reference [5] suggests that to drive a link the distance of the cube side-length a pump power of 50 mW is needed, which is considered a "low power" laser. An experimental Raman continuous wave laser has a lasing threshold of 26 mW and an output power of 50 mW, so in the lab at least a micro-laser with the needed power has been produced. [6]

If a signal is "dropped" via a ring resonator the signal loses power. A 50 mW pump signal, after being dropped, will have about 100 times less power: 500 μW. This attenuated signal would be applied to a Bragg grating resonator as described in [5] and pictured below.

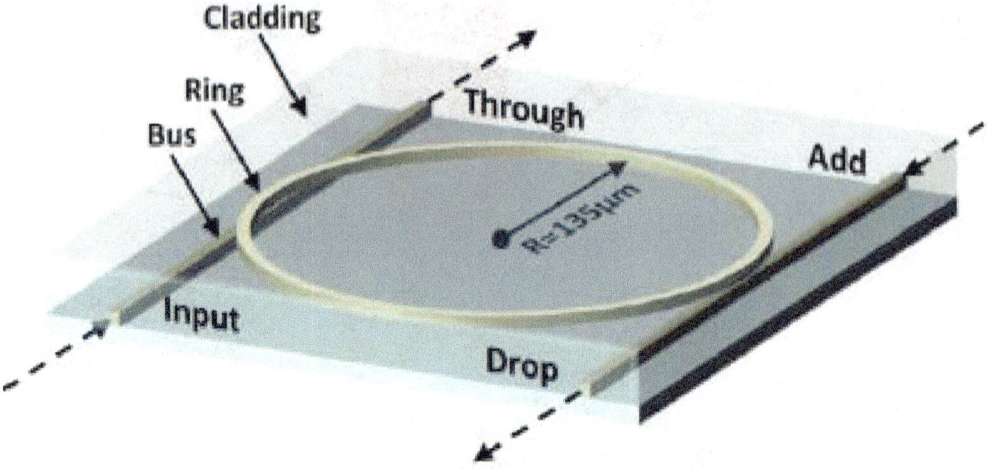

(courtesy [5])

The authors of [4] note that the laser power can be decreased by introducing irregularities in the silicon, which has the effect of increasing absorption. Designs of this character are called *super-absorbers*.

5.7. Conclusion
This chapter has taken a close look at the optical data paths that are a critical part of the display device design. By adding some functionality in each column header / controller, as well as a phase-locked loop station at the foot of each optical plane, super-structured Bragg fibre gratings can be dispensed with in the body of the cube. They will still be needed at part of the landing pad circuitry after the various clocks have been recovered.

To support data routing, NRZ encoding and clock recovery has been presented, along with an experimental silicon-grating laser coupling solution for the modulated data at the optical receiving station. Various timing and power considerations have also been presented.

References
[1] Poon et al. *Large-scale-integrated silicon photonics using microdisk and microring resonators*, 2009, online
[2] Li et al. *All-optical passive clock extraction of 40 Gbit/s NRZ data using narrow-band filtering*, Optics Express, v. 12 no. 14, 2004, online
[3] Singh, Anjali. *Modulation Formats for High-Speed, Long-Haul Fiber Optic Communication Systems*, white paper, Inphi Corporation, 2001?, online
[4] Caspers et al. *Ultrafast silicon-based active plasmonics at telecom wavelengths*, Optics Express, v. 18 no. 19, 2010, online
[5] Moss et al. *CMOS Compatible Micro-Ring Resonator Lasers*, IEEE conference paper, 2013, online
[6] Rong et al. *Low-threshold continuous-wave Raman silicon laser*, Nature Photonics, v. 1 no. 4, 2007

Chapter 6

Signal Path

6.0. Introduction

Here we look closely at optical interconnect design issues.

6.1. Optical Signals

Recall that the cube's *modus operandi* is optical interconnects between the backplane and the cube. In the simplest design the backplane contains 128 cache memory pin-outs, which in turn are routed to an optical modulator configuration, each line driven by a nominal 50 mW pump laser and a modulator of a serial digital signal. 128 lasers form a laser strip which is the interface between the backplane and cube.

Each signal is routed to a parabolic concave circular contact that mates with a convex circular contact on the cube. All of the contacts are situated along the lower edge of the cube that's seated by gravity in its cradle. This design allows the cube to be easily removed from the backplane to be appreciated as an aesthetic object.

This optical interconnect design facilitates high-speed data transfer between the backplane and the *base plane* of the cube. Each cache pin corresponds to an *optical plane* in the cube; there are 128 optical planes, one for each pin position. The logic design for one optical plane applies to all of the other planes. The planes are updated in parallel.

There are several concerns with an optical data path, including: a) the type of waveguide used; b) the distances traveled; c) the non-ideal effects associated with the waveguide (parasitical resistance loss combined with a waveguide skin effect; dielectric loss; reflections because of impedance mismatch; channel coupling; and electromagnetic radiation loss); d) the speed of the data signal; e) the voltage levels of the signal; and f) whether the signal is differential or single-ended.

An optical receiver design starts with a *transimpedance amplifier* (TIA) that generates voltages in the millivolts. This front-end is followed by a *main* or *limiting amplifier* configuration that boosts the voltage to digital levels. A clock rate of 10 GHz can be achieved with CMOS technology [1, 2]. An optical data path can span kilometers, or span two chips. This latter is called a *chip-to-chip* interconnection, and this distance is on par with the distances in the display device.

All of the non-ideal effects listed above affect the signal-to-noise ratio of any receiver. As well, these effects can affect the transfer signal's amplitude and cause phase distortion. An equalizer is therefore employed to compensate for these effects.

The graphic below shows a signal on the receiver end without an equalizer (left), and with (right). The diagram on the right is called an *eye diagram*, so named because of the eye-like shape of the enclosed space. The eye should remain open to avoid bit errors.

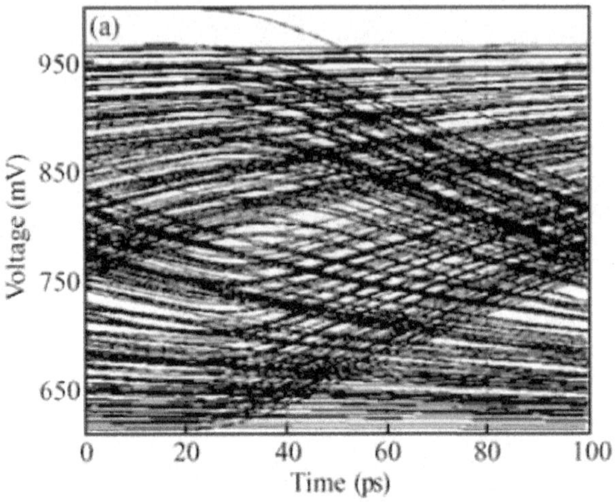

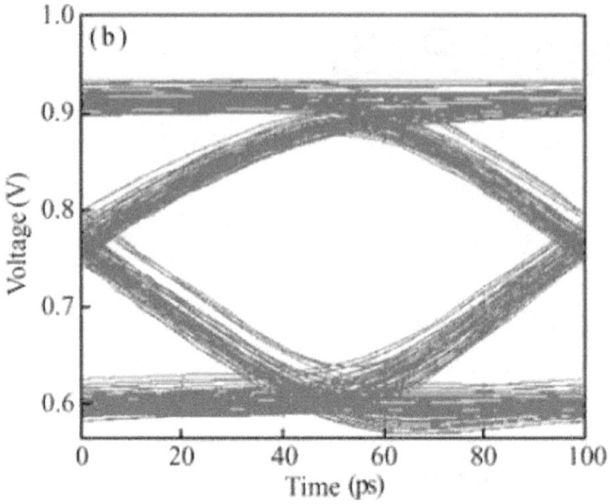

(courtesy [2])

The eye diagram is a good indicator of the integrity of the optical data stream. For the display device, the high-speed data should be limited to the *base plane* of the cube, which is like a micro-motherboard. The CMOS receiver designed by Zicha [1] is only the analog front-end because it contains no clock extraction or equalization circuitry. It contains a photo-diode which detects the optical signal, the TIA, the main gain stage, and output buffers to a digital trace.

The CMOS solution designed by Zhuo *et al* [2] is high speed but non-optical. It can be called the "3-by-5 chip" because the entire solution fits on a chip with dimensions of 300 μm x 500 μm.

The 3-by-5 chip is designed for chip-to-chip interconnects; there's no photo-diode and transimpedance amplifier, nor a gain stage. The design assumes that the signal is arriving on an FR4 PCB trace, which is a low-cost copper on dielectric solution. For this reason the design is called a *wire-line* receiver. The solution is all-in-one because it contains:

1. a clock data recovery (CDR) circuit;
2. a programmable continuous time linear equalizer (CTLE); and
3. a decision feedback equalizer (DFE).

The CDR includes a phase-locked loop (PLL) circuit and a voltage-controlled oscillator (VCO). The Zhuo *et al* design is quite sophisticated: first because the team managed to make these many different elements work exceptionally well together; and second because they implemented a half-rate-period calibration CDR with a *calibration enable.* The calibration enable is only triggered when a *pseudo-error* occurs – that is, when various phase conditions are present that might lead to an error.

The CDR includes a phase detector (PD) with memory – it holds the most recent effective phase results. In effect, part of the CDR can go into a power-saving sleep state until the next round of calibration is required. The combination of all of these features results in a bit-error ratio (BER) of less than 10^-12. As Zicha notes, a BER less than 10^-12 is considered to be "error free."

This is all the more remarkable because the chip doesn't depend on data scrambling. Data scrambling refers to encoding the binary data to ensure that there are enough transitions to keep the phase-locked loop in line, which in turn is related to the effects of equalizing the data signal on the receiver end. An equalizer can introduce *baseline wandering* because the high-pass filter behaves badly when the data frequency is reduced by *data idling*, when a long sequence of 1's or 0's are transmitted. One solution is to use FDDI encoding to make the data sequence *transition rich*. This kind of stream is also referred to as DC-balanced.

Encoding and decoding circuitry increases the silicon footprint, creates latency and dissipates power. Ideally we'd like to do without it, especially since NRZ data encoding is not DC-balanced. Ideally the equalizer circuit should compensate for baseline wandering.

Consider the image below.

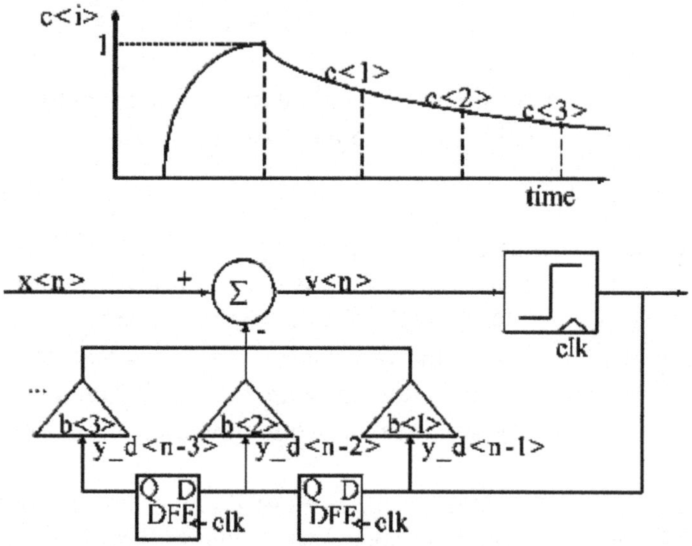

(courtesy [2])

The top part shows the band-limited response where *c* represents inter-symbol interference (ISI). ISI arises when the receiver bandwidth is narrower than that of the incoming data. This causes signal distortion, where for certain bit sequences the signal swing is reduced. ISI also leads to the wandering baseline problem. The bottom part of the diagram shows the decision feedback equalizer (DFE) system. The DFE is a *quantized feedback* circuit because the output signal is fed back to the input. The circuit restores low-frequency components in the data signal and maintains a low error rate.

In general a feed-forward design compensates for channel distortion in the form of the low pass channel's frequency-dependent loss. The coefficients *b* are programmable and can cover a wide bandwidth. That is, if the CTLE has the transfer function CTLE(*s*), and the transfer function of the channel is C(*s*), then

C(s) X CTLE(s) = constant.

Referring to the diagram above, given an ideal input signal *s*, the received signal *x* can be expressed as:

$$\begin{aligned} x<n> = {} & 1 \times s<n> + c<1> \times s<n-1> \\ & + c<2> \times s<n-2> \\ & + c<3> \times s<n-3> + \ldots \end{aligned}$$

The interference *c* and the resulting signal swing causes the received signal *x* to suffer. The voltage output signal *y* of the DFE can be expressed as:

$$\begin{aligned} y<n> = {} & x<n> - b<1> \times y\_d<n-1> \\ & - b<2> \times y\_d<n-2> \\ & - b<3> \times y\_d<n-3> - \ldots \end{aligned}$$

where *b* is a programmable coefficient that's supposed to compensate for the signal distortion. Substitution of *x* from the first equation into the second gives:

$$\begin{aligned} y<n> = {} & s<n> + (c<1> \times s<n-1> - b<1> \times y\_d<n-1>) \\ & + (c<2> \times s<n-2> - b<2> \times y\_d<n-2>) \\ & + (c<3> \times s<n-3> - b<3> \times y\_d<n-3>) + \ldots \end{aligned}$$

This equation minimizes the error between input *s* and output *y*. The circuit is said to "remove the post-cursor inter-symbol interference." The DFE used in the 3-by-5 chip has "order three" if one thinks of a polynomial equation.

Moving on, the core power usage of the 3-by-5 chip is only 39 mW (29 mW for the CDR, and 10 mW for the combined CTLE + DFE). Given 128 receivers in the cube, all lined up along the laser strip, the total power consumption is 39 mW x 128 = 4.99 watts. On the backplane side the laser pump requires 50 mW, and given a straight design of 128 separate microring lasers, this adds 50 mW x 128 = 6.4 watts. Thus the major power use between the backplane and cube amounts to 11.39 watts. Keep in mind that the final design only uses 32 optical couplers. This reduces the total power consumption to 2.84 watts.

There would be some additional circuitry along the lines of Zicha's analog front-end to convert the light signal into a 1 volt CMOS signal. It should be clear by now that the receiver's role is to clean up the signal through equalization, and to perform clock data recovery. The resonator design for clock extraction featured in Chapter 5 only served as a comparator; the proposal was to implement the phase-locked loop in a separate CMOS chip. The 3-by-5 chip (shown below) is exactly that chip; since it has its own PLL it supercedes and improves upon the resonator design by a wide margin. All in all, the work of Zhuo *et al* is a remarkable achievement.

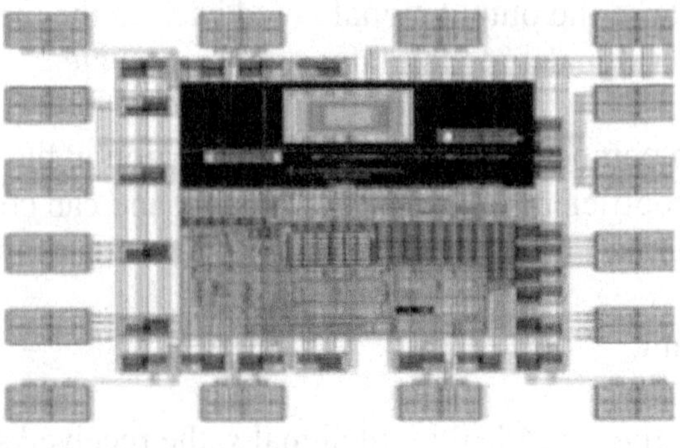

(courtesy [2])

6.2. Transmitter

The receiver on the cube side is complemented by a transmitter on the backplane side. The transmitter must light-modulate a digital signal. However, the serial data stream that feeds the light modulator must be created in advance, and this is done with a multiplexer. The following circuit is from the paper by Tanabe *et al* [3].

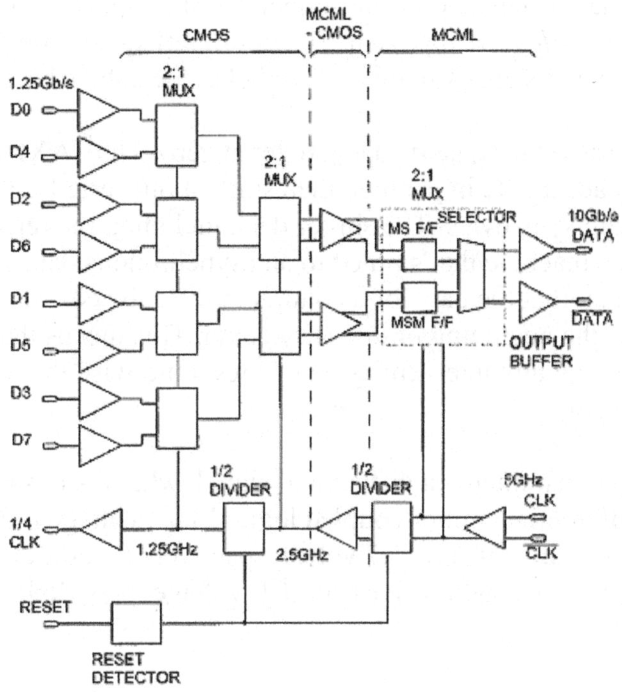

(courtesy [3])

The divided clock will progressively combine the individual bits into a serial stream that's output at 10 Gbits / sec. The circuit uses both edges of the clock signals, so only a 5 GHz clock is needed. This circuit uses MOS current mode logic (MCML) which is faster than CMOS. It's used on the fastest clock, while regular CMOS is used on the slower clocks. This circuit is a power hog at 112 mW for the 10 Gb/s speed. This is an 8:1 MUX. Three 8-bit data words could be clock-divided with this circuit, corresponding to 8-bit values for red, green, and blue.

6.3. Cascaded Multi-Drop

Recall that Bragg gratings only reside in the cube's base plane. Since the smallest write size is all *n* elements of a voxel column, a *multi-drop* solution can be used to address each level of each column.

Since the data is encapsulated in the NRZ data packet for the entire column, and since the data stream is serial, no two levels will be latched at the same time. The penalty of having to write an entire column to output a single voxel (for single writes) isn't large, since the data speed is fast enough for the additional writes to be transparent. In essence, one write for an entire column makes the display cube two-dimensional (more on this later). The full-column versus individual voxel tradeoff is only an issue when addressing individual voxels. In *refresh mode* one would update the entire column no matter what.

This design tradeoff means that the gating signal can be cascaded up the column. Ideally this would involve one line that daisy chains from one level to the next. This kind of thing has been done before, a good example being the Transparent Asynchronous Xmitter-Receiver Interface or TAXIchip from AMD [4]. The general interface is *point-to-point*, but in cascade mode the interface is a *multi-drop*.

Suppose a counter clocked to the NRZ data stream feeds a multiplexer, and the multiplexer selects one of 128 line-outs. In that scenario single bits are decoded for each level in turn, until all nine bits (three bits per red, green, and blue) have been latched for each level. This is an *interleaved* approach. The TAXIchip approach is similar, but the data is latched for the equivalent of all nine bits for a level before moving on to the next level. This is a *non-interleaved* approach. The non-interleaved ordering makes sense for the display device because all nine bits are available in parallel at the cube's base plane for any one level.

Recall that a sync bit at the packet head starts the counter at zero. The TAXIchip also uses a sync signal. The TAXIchip transmitter loads a byte in parallel, then shifts it out serially. The idle time between transmitted bytes is filled with sync bytes. The shifter data, including the repeating sync pattern, is generated by an internal state machine that's timed to an asynchronous data strobe signal, or to a clock signal. However, it's important that the data for a column in the cube contains no intervening sync bytes, because the sync byte marks the beginning of the daisy chain. This means that all 128 data bytes must be shifted out continuously, without any intervening sync bytes, otherwise the data when it's received at each level will suffer a *framing error*.

The output of the TAXIchip transmitter is a differential signal, which is to say, it's not a long-haul optical-level voltage but rather a differential emitter-coupled logic (ECL) voltage referenced to +5 volts. High and low ECL *serial output voltages* are –.9 and –1.7 volts respectively (average). This is very important since the TAXIchip architecture transmits and receives an ECL voltage, not a light signal. AMD calls this a *media interface*.

The transmitter chip is matched to one or more receiver chips. Each *level* in the display device can be compared to a *receiver* in the TAXIchip system, and the first receiver in the cascade is the *primary receiver*. (Note that the name *receiver* shouldn't be confused with an *optical* receiver. Here the receiver is non-optical.) Each receiver has an IGM ("I got mine") *output* tied to the next receiver's CNB ("catch next byte") *input*. The primary receiver's CNB is tied high; the tail-end receiver's IMG output isn't connected, since it's the last in the chain.

The beauty of the AMD design is that the propagation time from receiver to receiver is covered by the time it takes to read in 8 bits for each receiver. The non-interleaved data sequence provides the delay needed to propagate the IMG/CNB signal. The IMG signal goes high in the middle of the 8-bit serial segment, providing plenty of time for the next receiver to prime itself for the next 8 bits.

Each receiver catches a valid byte and then holds it. The *primary receiver* ignores all ensuing data until it recognizes a sync signal, which causes its IGM output to go low. All other receivers down the line ignore the data until each receiver's CNB input goes low and then high again. The cascade starts with the sync signal at the primary receiver, so there must be a sync signal at the head of every data packet. There is one packet for each column write.

The idea is illustrated below for three receivers. Notice that the sync byte causes all receivers to essentially reset themselves. This takes the form of the IGM/CNB link signal going low and propagating itself with a slight delay from receiver to receiver.

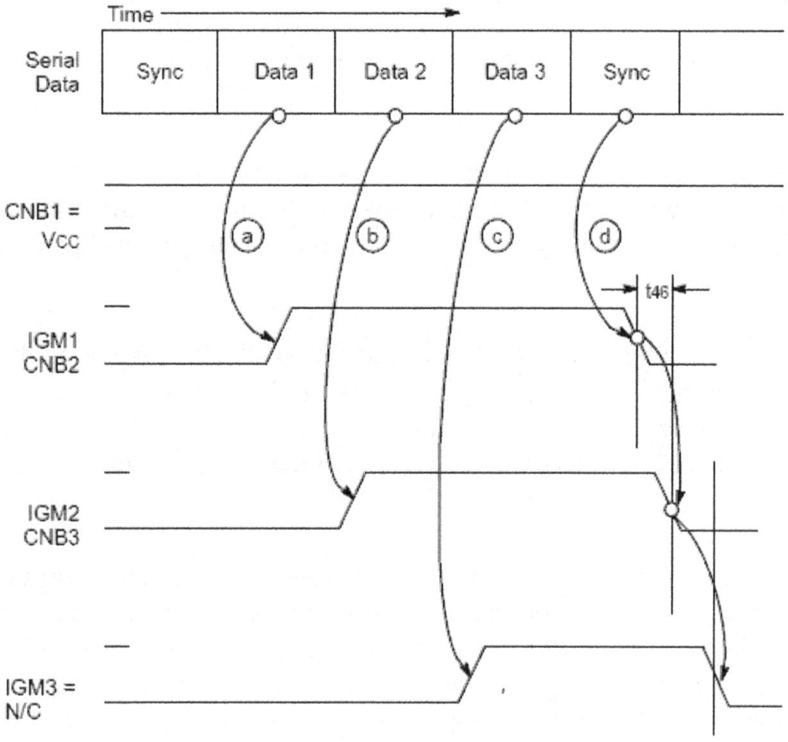

(courtesy [4])

Therefore the primary purpose of the sync signal is to restart the cascade circuitry. Outside of the cascade logic, buffer logic recognizes the sync signal and moves data stored in the receiver to the drop port. In this regard (but only in this regard) every receiver recognizes the sync byte and internally transfers the data bits from an input latch to a decoder latch, and after another clock cycle to an output latch, where it can be picked up by the external drop port circuitry when a data strobe signal goes high.

Every receiver drops its data in parallel. Since sync bytes are sent during idle, an internal latch count should be checked: if it's zero, there's nothing to drop. In any case, this parallelism is meant to accomodate a wide data bus – which is why the cascade feature is offered in the first place. For example, four chips can be cascaded to make a 32-bit wide data bus. For the display device the cascade is used to latch voxel data, and since each voxel is independent the simultaneity of the drop isn't critical.

The AMD documentation notes that the minimum number of syncs required in a user data stream is dictated by the system sensitivity to running errors and the system's built in error detection mechanism. As we will see in the next section, the timing is relaxed for column writes and so errors will decrease to virtually zero. AMD advises that it's a good practice to send a sync every 1000 bytes. This limit is well above the 128 bytes (nominal) that the display device will be sending between syncs.

6.4. Clock Relaxation and Column Signals

The receiver design in section one makes it clear that the sooner the optical signal is dispensed with the better. From a timing point of view the sheer speed of the signal may make the IGM/CNB cascade difficult to implement. Ideally a slower clock on column data would be desirable.

The solution to this dilemma lies in the 128 column headers / controllers. Since the dimension in column-space is the same as the dimension in level-space, the column headers can be used to shift through groups of nine bytes, and at the same time slow down the transfer speed. As already mentioned, the transfer rate of the display cube is 1000 times slower than the high-speed 10 Gb/s data lines being handled, for example, by the 3-by-5 receiver chip.

Therefore, there's plenty of time between column writes on the fiber-optic lines to latch data in the base plane, and divide the clock down to something that makes sense for the transfer of the serial stream to the levels, assuming that data can be retrieved from the cache memory in the backplane and transfered at high-speed without data gaps.

One can imagine a fast clock, being the NRZ clock; and a slow clock, being something divided to the megahertz range. The difference between the optical speed and the column write speed means that the display device operates in *burst mode* rather than continuous mode. Each burst is a column write of 128 voxels. The fast data is buffered at the column headers, and then gated to the selected column at the slower clock speed.

This way, the column cascade can be toned down a bit to a more manageable (digital) speed. Each column header would be subjected to a base-plane cascade, similar to the column cascade, except that the primary receiver would be the selected column and there would be no parallel data drop. The base-plane cascade would partition bytes at high speed using all 128 column headers, which can also be called *base stations*. For example, there might be a 9-bit shift-in register for each column header. The data would be shifted in at high speed; but shifted out with a slow clock onto a slow serial bus that runs parallel to all of the column headers. Then, the selected column would gate through the bus data to the levels for that column on the slow clock.

By the time all of the shift registers have been unloaded, the column headers will be ready for the next high-speed burst (e.g. the next column write). Note that each sequence of 9-bit words is transmitted from the backplane in a continuous stream; since all 9 bits are available in parallel from different cache slices, and since they need to be serialized, then while serialization is taking place the next level can be fetched from cache memory in a concurrent operation that hopefully will result in no data gaps.

This design may seem overly complicated, but its main benefit is to isolate all of the high-speed ringing and reflections and bouncing noise to the base plane. By the time the data is ready to be sent up a column, it will have been cleanly captured, and can be cleanly re-latched at every voxel at a slower clock rate. The secondary benefit is that the data can be transmitted via a metal (digital) connection rather than optical, which simplifies the latching circuitry at each voxel.

In general, 3D topologies for optical switching are very hard to pull off. If one thinks of the optical plane as a 2D topology, even this is hard to do. This double cascade scheme limits the high-speed operations to a 1D (linear) topology.

As well, this approach is more in line with the TAXIchip architecture which is TTL/ECL based, and facilitates quick picks from the (digital) data stream without the overhead of optical conversion.

Note that the clock is recovered at the clock recovery station (CRS), which now includes equalization, the whole kaboodle in the form of something like the 3-by-5 receiver chip. Instead of decoding the column address header optically with super-structured fibre Bragg gratings, a simple "and" gate array can do it digitally. The column address header is still a required data item, but once past the CRS everything turns digital.

Note that, as per the TAXIchip architecture, each voxel circuit must count bits. The byte latching might be in the form of a simple clock counter and shift register.

This brings us to the various lines that are cloaked in each column. The lines are:

1. Lower speed digital serial data bus (megahertz range).
2. Lower speed clock signal.
3. IGM/CNB interconnect (one line cascaded from level to level).
4. Power (for the LED circuits, CMOS).
5. Ground

The serial data bus voltages can be at emitter-coupled logic (ECL) levels, but the signals won't be optical, they'll be electrical. Meanwhile, the base-plane high-speed bus will employ some kind of microstrip compatible with the 3-by-5 chip's 1 volt peak-to-peak signal output.

6.5. Conclusion

This chapter has presented the design tradeoffs of optical coupling and its downstream extensions. The design is integrated by combining clock extraction with signal equalization, but is slightly complexified (though ultimately simplified) with the decision to limit high-speed circuits to the cube's base plane, which acts as a micro-motherboard. A cascade scheme is one way to drop the data at each of the cube's levels for a column.

References

[1] Zicha, Nicholas. *High-Speed Optical Receivers in Nanometer CMOS*, Masters Thesis, Department of Electrical & Computer Engineering, McGill University, 2009, online
[2] Zhuo et al. *A 10 Gb/s receiver with half rate period calibration CDR and CTLE/DFE combiner*, Journal of Semiconductors, v. 30 no. 4, 2009
[3] Tanabe et al. *0.18-µm CMOS 10-Gb/s Multiplexer/Demultiplexer ICs Using Current Mode Logic with Tolerance to Threshold Voltage Fluctuation*, IEEE Journal of Solid-State Circuits, v. 36 no. 6, 2001, online
[4] AMD Am7968/Am7969, *TAXIchip Integrated Circuits, Data Sheet and Technical Manual*, 1994, online

Chapter 7

Nano Components

7.0. Introduction
Here we take a plunge into the ocean of nano-technology.

7.1. Waveguiding at the Nano Scale
The backplane interfaces via a *laser strip* aligned with one edge of the cube. Since the base plane resides within the cube proper it has the benefit of being able to guide optical signals to the rest of the cube directly, assuming the signal has arrived cleanly from the laser strip. As outlined in the previous chapter, however, the sooner the optical signal is converted to a digital signal in the base plane the better.

The upstream conversion to digital is desirable not only from a bandwidth perspective, but because, as outlined in the paper by Campenhout *et al* [1], the power of a microring laser is limited. The picture below shows their experimental setup.

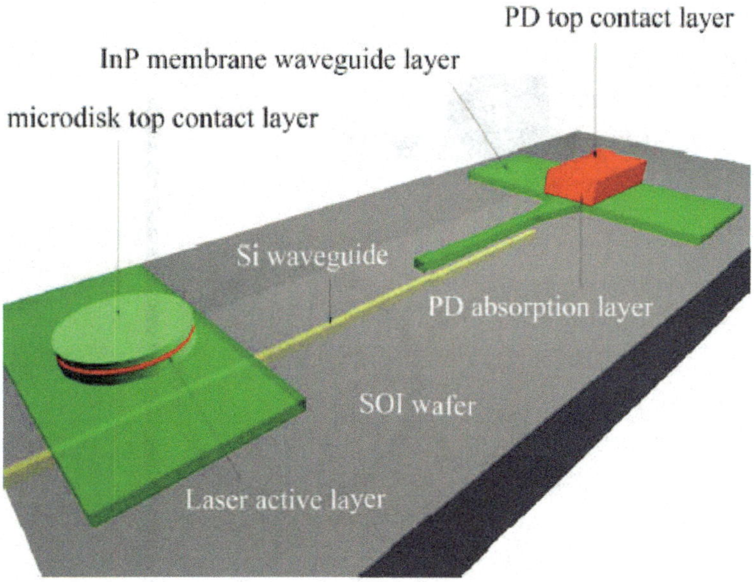

(courtesy [1])

The length of the waveguide link as shown above is 7 millimeters – far short of the cube side length of 50.8 mm. But 7 mm may well accomodate the distance from the photonic integrated circuit (PIC) in the backplane to the optical coupler. In other words, once the optical signal is received by the cube, it's immediately converted to a high-speed digital signal. The driving power of the microring laser is given as 300 nW for pulsed mode, and 1.6 μW for continuous mode. The microdisk laser has a diameter of 7.5 μm and a thickness of 500 nm – well within the boundaries of the 400 micrometer side length of each nominal optical plane of the cube. Which is to say, the laser is extremely small.

As shown above, a photo-detector (PD) is able to detect pulses sent over the waveguide, but only if the laser current isn't too high. In general, nano-component tolerances must be closely matched.

7.2. Micro-Lasers

In the nano-vernacular a *microring* isn't necessarily a *micro-laser*. It's a *resonance* device with lasing capabilities. A good overview of micro-resonators is given in [2]. Optical resonators can be fabricated by exploiting either: a) almost total internal reflection (ATIR) of light at the interface between a dielectric material and the surrounding air; or b) distributed Bragg reflection (DBR) from periodic structures such as multilayered structures or arrays of holes. A resonant hole array was explored in terms of an LED light-amplifier in Chapter 2.

A *microring* resonator is a versatile and popular choice among nano-engineers. Other types include a *micro-torus* resonator (which has a high Q or quality value); or a variation on the microring called a *racetrack* resonator (shaped like a horse racing track). The Q-factor is a measure of the resonator capacity to circulate and store light, and is usually defined as the ratio of the energy stored to the energy dissipated. The *free spectral range* or FSR is the spacing between neighboring high-Q resonances.

7.3. Microring Modulator

A microring modulator for optical pulse generation has been designed and built by a group from Cornell [3]. The microring is 12 μm in diameter, shown below.

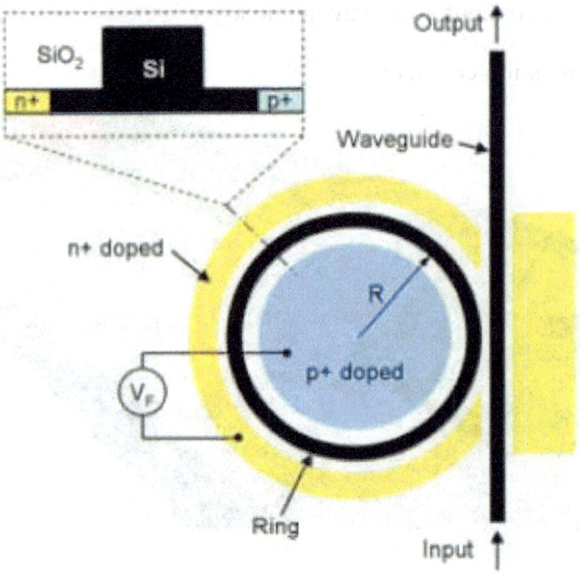

(courtesy [3])

The P-i-N cross section of the microring is shown in the upper left corner. The modulator operates by bipolar charge injection and pulse extraction using the microring. The incoming wavetrain is converted to an *emphasized rising edge* waveform which reduces the energy required per bit while removing pattern-dependent jitter in the modulated signal. This is shown below.

(from [3])

One measure of the quality of the resonance is the *extinction ratio*. For the modulator a higher extinction ratio is more desirable since otherwise photons interfere with the high-frequency signal. The system has an extinction ratio of 8 dB at 16 Gbps; and 3 dB at 14 Gbps. The degradation due to photons lingering in the resonator cavity can be treated by lowering the Q value of the cavity.

In any case the system works on silicon and is small enough to do the job for data arriving from the pin-outs of display cache memory. Recall that this data is serialized before being modulated for optical transfer.

7.4. Resonance

As hinted at in the previous section, cavity resonance is the main factor for getting these nano-components to work. The work by Che *et al* [4] simulates the guided emissions for cavity configurations with 2, 3, and 4 waveguide ports, as shown below.

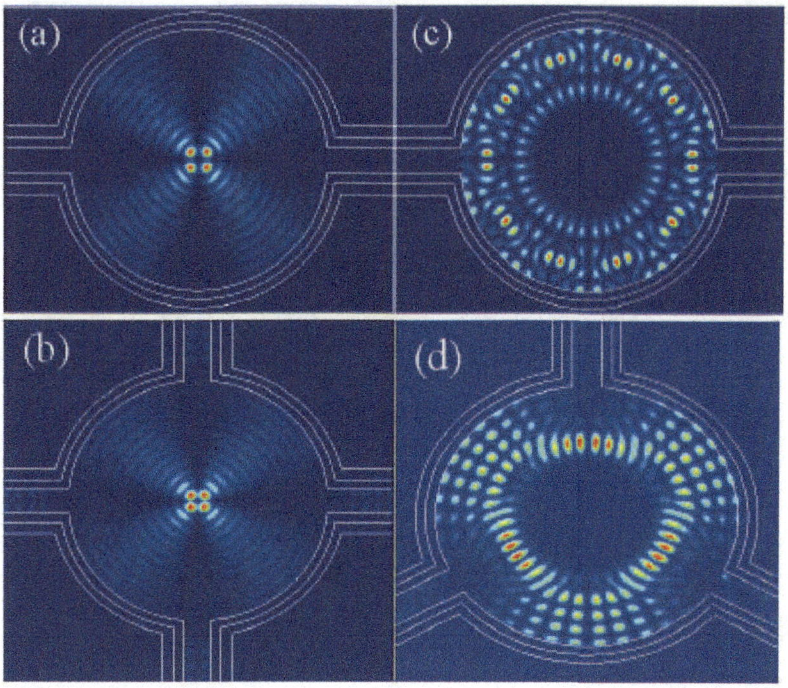

(courtesy [4])

These simulations apply to a transverse magnetic (TM) polarization of the microring. In general there are two orthogonal polarizations of the electromagnetic radiation inside of the cavity. A *mode* corresponds to a standing wave pattern in the cavity. A *mode number* in some dimension is defined as the periodicity of the wave pattern in that dimension. The modes that develop between a metallo-dielectric confined microdisk and wide waveguides are dominated by *coupled* modes and *high-order* modes. Box (a) above shows a high-order mode, TM(2, 12), designated by the azimuthal and radial numbers (m, l). The periodicity in the radial direction is 12; and in the azimuthal dimension, e.g. the rotation angle in the plane around a central axis projecting out of the page, there is one division of the wave pattern. The same mode is shown for a four-port configuration in (b). The energy focuses at the center of the microdisks. Boxes (c) and (d) show coupled modes, which have a zero energy intensity value near the center. The three-port configuration in (d) has a triangular-shaped interference pattern. The numbers for (c) are a coupling of TM(18, 5) and TM(28, 2) – you can count 9 spots in each quadrant of the inner ring, and 28 bright spots on the outer ring; and for (d) the coupling is TM(18,5) and TM(21, 4).

As a rule the introduction of additional waveguide ports increases the power loss out of the ports in the various modes. Some modes survive the port layouts, others don't. The mode quality-factor Q has the best response when there are 2 or 4 ports, and degrades above a certain port width when there are 3 ports. The idea is to drive 2 to 4 opto-electronic elements with a single light source (a single-mode laser with selective pumping) via the output ports as shown above. This scheme is instructive but since the same signal is sent to all ports, it doesn't really apply for the cube. Each optical port of the cube carries unique light data, and so requires a dedicated laser to drive the signal.

7.5. Optical Flip Flop

Microring resonance can be made to switch direction, and this bidirectionality can in turn be used to make an optical flip-flop. A flip-flop represents one bit of an optical register. Some kind of register (optical or electronic) will store an aggregated value that is subsequently converted to an analog signal by the LED circuitry. An optical register stores data in a *very* small footprint. The system of Liu *et al* [5] is shown below.

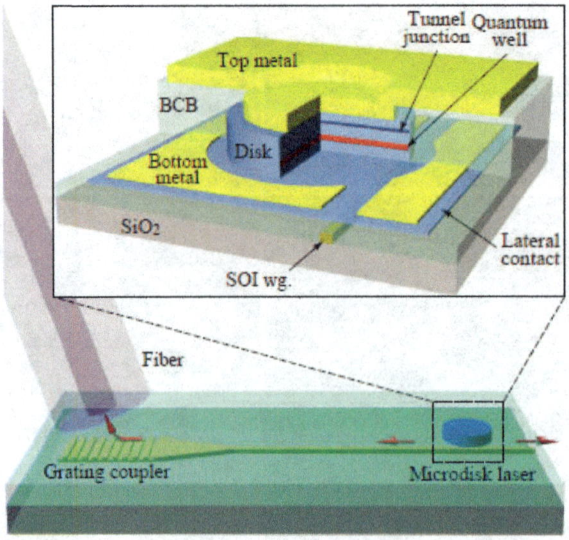

(courtesy [5])

The microdisk laser has a diameter of 7.5 μm. Note that the fiber optic cable and the grating coupler illustrated above is for testing purposes only. The authors used III-V/SOI (silicon on insulator) heterogeneous integration to achieve good lasing performance on silicon. The technology is CMOS compatible. Nano-components tend to be CMOS compatible, an example being CMOS driven micro-LEDs. The type III heterostructure is also the proposed medium for the laser integration on the backplane.

Theoretically a microring has a whispering gallery mode (WGM) behavior – although this is contested by some researchers, who prefer a Schrodinger wave model. Given a large enough photon density in the WGMs, the laser has a bistable unidirectional behavior. For this bistable behavior to work well, the laser should be shielded from coupling noise due to reflections and feedback in the waveguide lines.

Continuous-wave light injected in a long pulse at 360 nW (the same power as for the waveguide link described in section 1) is enough to reliably switch the laser from a clockwise (CW) to a counter-clockwise (CCW) mode. Switching between these modes takes only 60 pico-seconds. The diagram below shows a set / reset sequence. The *set* signal arrives from the left, while the *reset* signal arrives from the right. Since the experimental setup had only one external coupling to the waveguide, the authors only tested the *set* capability.

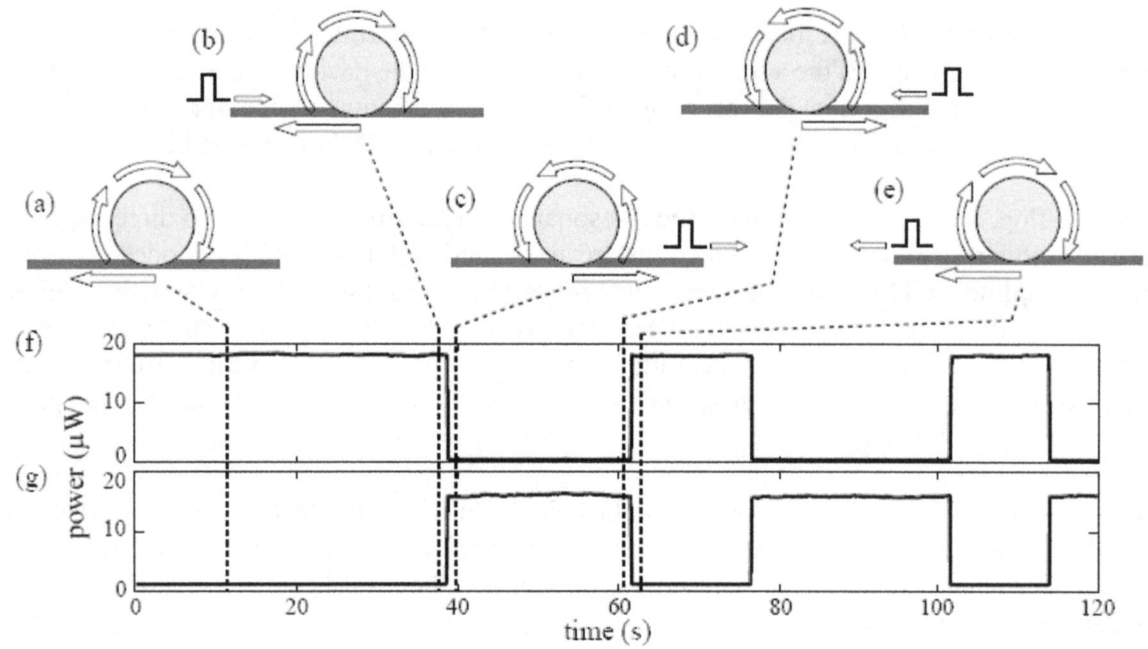

(courtesy [5])

7.6. Microring with Bragg Grating

A critical component of the packet switching paradigm outlined in Chapters 4 and 5 is the Bragg grating which serves as a discriminator or correlator for address encoding and decoding. Shi *et al* [6] have come up with a superb start on a nano-sized integration of a Bragg grating. Their design is shown below.

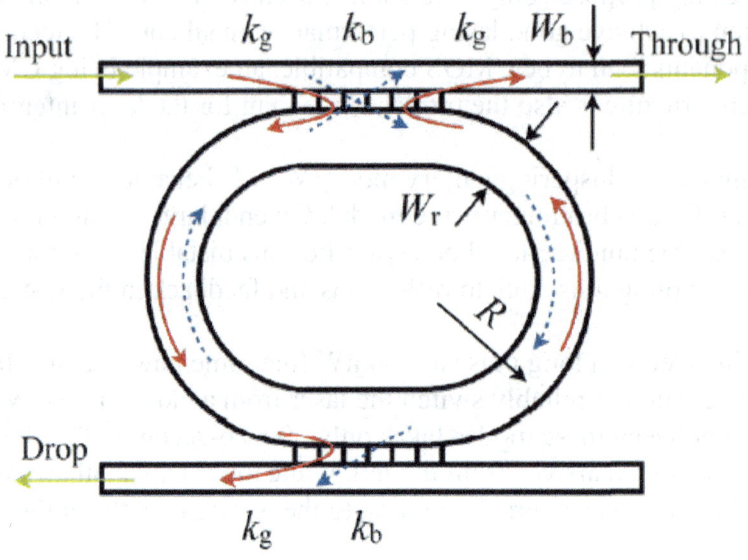

(courtesy [6])

The authors designed a racetrack resonator interfaced to two Bragg gratings (banded strips, above). If an incoming address packet matches the address of the receiving node the packet is *dropped* – i.e., dropped from the network to be processed by the receiving node. The design shown above implements a single-bit Bragg grating drop processor. This element like the others is SOI and CMOS compatible.

As with the flip-flop, a single input excites optical resonances in the ring cavity in two directions: clockwise and counter-clockwise. These directions are associated with two coupling modes: a) broadband *co*-directional coupling; and b) grating assisted *contra*-directional coupling – k_b and k_g in the diagram above. The ring oscillates primarily in the counter-clockwise direction associated with the grating-assisted coupling. This means that the broadband coupling is, by comparison, relatively weak. Efficient coupling between the two waveguides (input and drop) only occurs near the wavelength that satisfies the phase match, i.e. the Bragg grating signature.

The wavelengths at which the coupling occurs can be calculated with the effective indices of the waveguide ring and the waveguide input bus (top channel, above). Bragg reflection takes place inside the bus and strongly influences what's called the *through-port spectrum* (left, below). There's also a *drop-port spectrum* (right, below).

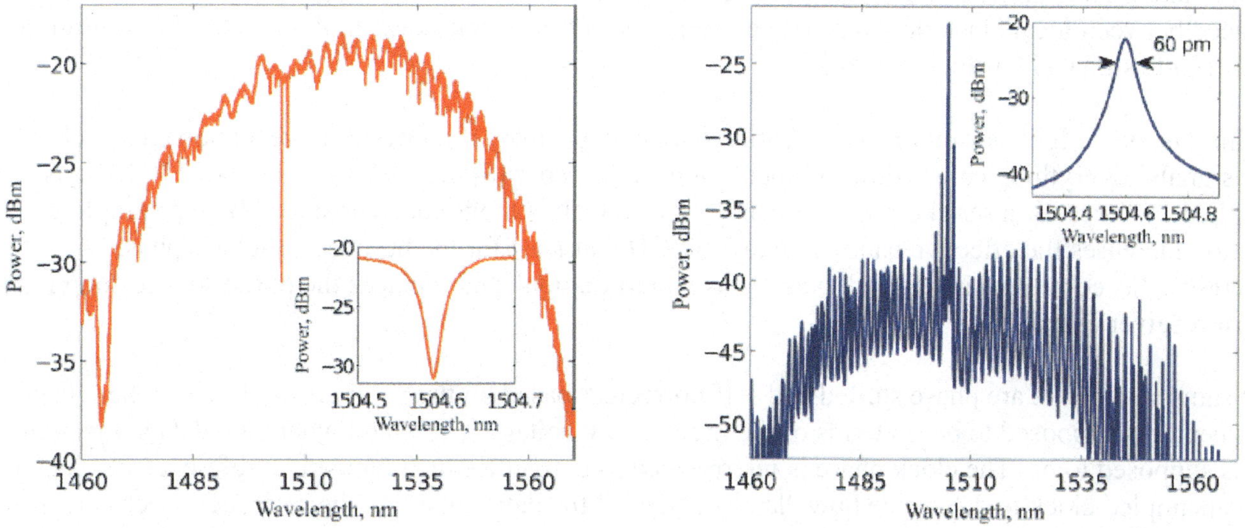

(courtesy [6])

Both spectra respond to the *dominant longitudinal mode* (or *selected* mode) which in this case is 1504.6 nm (insets). The drop-port spectrum (right) shows an out-of-band rejection ratio of 19 dB: compare the power of the red curve to the blue. The graphs show an excellent match between the input and the drop port responses.

Though an *n*-bit drop port could be used for data recovery, recall from Chapter 6 that the three-by-five chip takes a different approach. It uses a phase detector running on a four-phase clock and a decimated data oversampling clock to lock onto the input clock and decode the bits, which are converted from an analog to a digital stream. The digital stream is, for voxel-write purposes, further split into four signals using a second four-phase clock/data signal scheme.

7.7. Clocks

One display cube timing analysis has shown that the pixel clock has a very snappy period of 1.507 nanoseconds. The design goal is to slow the clock frequency down for transfer up a vertical conduit.

The clock can be slowed down by using a four-phase clock signal. The basic technology is outlined in the excellent papers by Jongsun Kim. [7][8] To do the conversion from parallel to serial on the sending side, and then to partly deserialize on the receiving end, a base frequency is "multiplied" (made faster). However phase drift, temperature drift, and various circuit nonlinearities can cause the multiplied signal to look very bad unless it's corrected. Normally, this is a what a "phase locked loop" achieves, but a frequency can also be multiplied with a multiplying delay-locked loop (MDLL).

MDLL is generically known as "injection-locked clocking" (ILC), which uses an "injection-locked oscillator" (ILO), and can be short-phrased as "injection locking." Logic blocks can be clocked by independent ILOs connected to a "low swing" global clock input. The clock signal amplitude doesn't need to be the full V_{dd} swing. Usually a full-swing signal from a central source needs to be buffered, and these amplitude buffers introduce jitter. Eliminating multiple levels of clock buffers on the "clock tree" reduces clock jitter. "Low swing" also characterizes high-speed logic such as CML.

Thanks to its high-speed and high-resolution phase correction, ILC provides deskewing of the output clock, within bounds. That is, the clock signal can't be sent too far from the deskewing center. Phase detectors can be placed between local clock domains to check and correct for clock skew. In this sense, ILC is used to pre-distribute signals in a clock network.

The thing is, one might as well throw the entire display cube away if it doesn't have a good clean set of clock signals. Everything comes down to those signals. In paper [8] the base frequency of 110 MHz is multiplied 16 times for a serial output frequency of 1.76 GHz. On the receiver side a 2x oversampling operation increases the effective bandwidth to 3.52 GHz, at least for the purposes of clock-phase correction. This correction is referred to as "phase interpolation" and replaces the phase-locked-loop found in most receiver circuits.

Eight sampling clocks are phase shifted and self-corrected, based on the 2x oversampling of where the data's center is supposed to be (in terms of the high or low voltage level), and where the data's transition edge is supposed to be. The clock phase is interpolated to correct for drift by using a reference clock and the oversampled clock to determine how "late" or "early" the data signal is. The reference clocks are also 8 in number and generated by the MDLL.

The maximum bit rate of 3.52 Gbps that applies to the incoming data stream is divided by 4 as the data is recovered in 4 data input receivers. Each of the 4 data outs is clocked by a phase-shifted 880 MHz signal, the phase shift being 90 degrees, as shown below.

Each of the 4 phase-shifted clocks has a shifted rising edge that serves to "clock in" a data bit. One can visualize the top signal trace as the serial stream, and the underlying traces as sampling edges. This 4-phase approach would be very good for eliminating crosstalk at the cube's base-plane switches.

The deserializer in [8] then goes on to make a 32-bit parallel word, but here Kim and the application of Kim's work to the cube design diverge. One way to cut down on the bit rate is to send not *one* serial stream up the vertical conduit, but *four*. Each stream corresponds to one region of a *cellular* voxel layout. Sending all four regions up in parallel allows each of the four lines to be clocked at one-quarter the reference clock period as determined in the timing analysis of Chapter 17, e.g. 755 MHz / 4 = 188.75 MHz.

In Jongsun Kim's application the clocks used to sample the data signal's "center" in the phase interpolation stage are also used to trigger counters in the deserializer. The 32-bit word is clocked out at 110 MHz, the math being four streams at 880 MHz divided by 8. This gives 4 x 8 = 32 bits. A 4:32 demux clocks out the data over 8 cycles of the 880 MHz clock. The 4:32 demux is arranged as 4 1:8 demuxes. The main function of the demuxing operation is to count in 8's – so it's mostly a counter.

Whether this four-phase separation of data is done as part of the "three-by-five" chip, or is applied to the fast serial data and serial clock that comes out in digitized form from the three-by-five chip, and is applied as its own processing island, is yet to be determined. There may be enough overlap in three-by-five phase detection and this four-phase post-separation for the features to be combined into a single processing unit.

One critical aspect of any MDLL is "lock in." Many MDLLs have a long phase-detecting period that causes longer locking time and even lock-in failures. The MDLL in [8] uses a phase detector that compares two inputs, clock-in and clock-out, and generates UP or DOWN signals depending on the phase difference of the two input clocks. The UP or DOWN signals go to a "charge pump" which generates a voltage which drives a voltage-controlled delay line (VCDL), which slows or accelerates the input clock (and resultant output clock). After 580 ns of this feedback loop the signal is "locked" and the clock output is normalized. Remember the output is a multiplied signal (1.76 GHz) based on a slower input clock (110 MHz).

In [7] a similar circuit, also by Jongsun Kim, extends the *power of two* multiplication {2, 4, 8} to programmable waveforms with fractional-ratio frequencies. In this case, N clock-out cycles are mapped to M clock-in cycles, where M can be {1, 2, 3}, and N {4, 5, 8, 10}. The "phase detecting period" is a time period during which the phase is adjusted, and over successive cycles the signal is "locked."

In terms of N and M, the phase detect lasts from the Nth falling edge of the clock-out, to the (M+1)th rising edge of the clock-in. There are three operation modes which shed further light on this timing:

sel1	sel0		VCDL mode
0	0	clock out	ring oscillator (RO)
0	1	clock in	reference injection (RI)
1	0	supply	supply injection (SI)

After lock, only RI and RO modes are repeated. The lock precludes the RO → SI and SI → RI mode transitions. The waveforms are shown below.

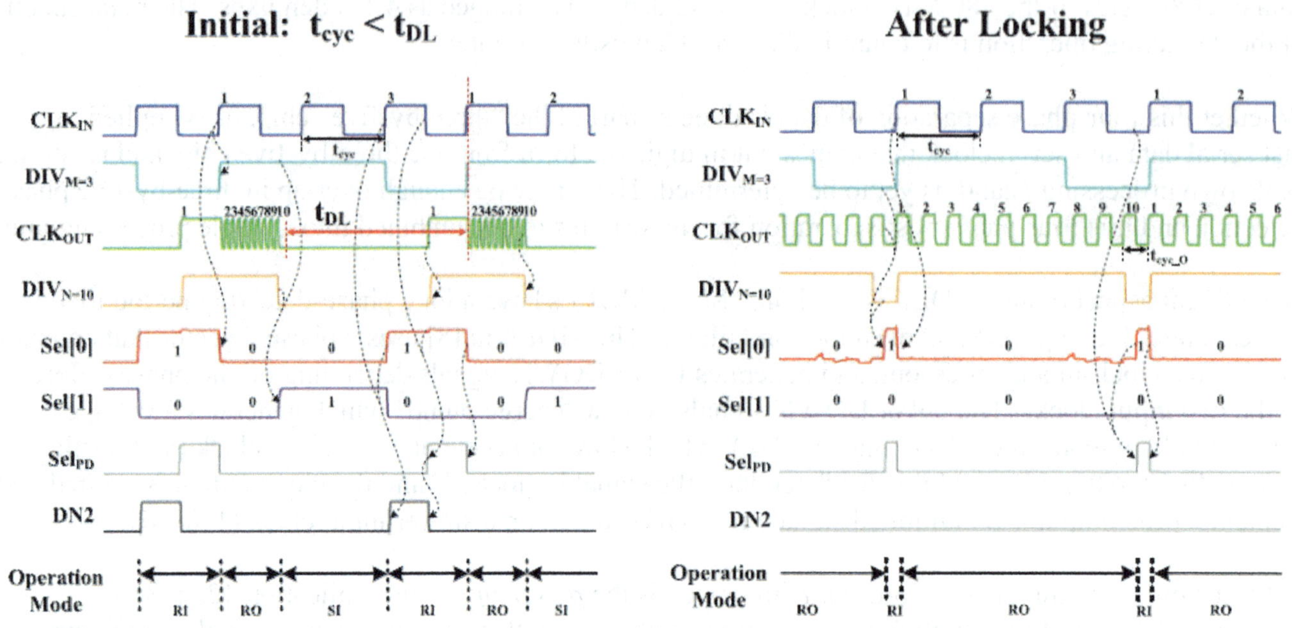

(a) (b)

(courtesy [7])

For the purposes of the display cube, the very fast serial data clock, running at 755 MHz, is divided by four, to 188.75 MHz. The period drops from 1.324 ns to 5.298 ns. But there is a further improvement to be made. The data for all four processing units is sent up the vertical conduit in parallel. This divides the fast clock by four and results in 4 serial lines instead of 1. So far so good. But recall that the data is sent to 128 levels. Why not send four parallel streams of four serial lines each to 32 levels each? This divides the clock by four again. The clock frequency for delivery up the vertical conduits becomes 47.1875 MHz, with a period of 21.192 ns. This means four times as much data must be prepared at the base plane, but the base plane is already operating in bursts and so it seems a fair tradeoff.

7.8. Conclusion

This chapter has presented a selection of cutting-edge nano-technology. Since the color planes need buffering circuits which reside outside of any invisibility cloaks, the prudent use of nano-circuits would reduce circuit footprints and increase transparency. The promise of nano-technology is its small size: once the manufacturing costs are absorbed, the material costs should be very low.

As well, a first look at signal timing and signal separation has been presented in terms of clocks and phase locked loop circuits. Optical packets received from the backplane can be decoded and separated into parallel streams to make the base-plane signal bandwidths more manageable.

References

[1] Campenhout et al. *Low-Footprint Optical Interconnect on an SOI Chip Through Heterogeneous Integration of InP-Based Microdisk Lasers and Microdetectors*, IEEE Photonics Technology Letters, v. 21 no. 8, 2009, online

[2] Benson et al. *Micro-Optical Resonators for Microlasers and Integrated Optoelectronics*, 2006, online

[3] Manipatruni et al. *High Speed Carrier Injection 18 Gb/s Silicon Micro-ring Electro-optic Modulator*, 2007, online
[4] Che et al. *Metallically confined microdisks with in-plane multiple guided emissions*, Optics Express, v. 19 no. 19, 2011, online
[5] Liu et al. *An Ultra-Small, Low Power All-Optical Flip-flop Memory on a Silicon Chip*, preprint, 2009, online
[6] Shi et al. *Grating-coupled silicon microring resonators*, Applied Physics Letter 100, 2012, online
[7] Kim, Jongsun, and Bae. *A 2 – 4 GHz fast-locking frequency multiplying delay-locked loop*, IEICE Electronics Express, v. 14, n. 2, 2017, online
[8] Kim, Jongsun, and Shin, Hyungsik. *A Low-power 3.52 Gbps SerDes with a MDLL Frequency Multiplier for High-Speed On-chip Networks*, Journal of Semiconductor Technology and Science, Dec. 2018, online

Chapter 8
Power Management

8.0. Introduction
Here we tackle the subject of power management and introduce a new feature of the display cube.

8.1. Display Cube Handling Mode
Previous chapters have outlined the design and operation of the backplane and the cube. The interface between the two takes the form of a laser strip aligned with one lower edge of the cube.

Originally the backplane communicated with the display cube through thousands of circular light contacts laid out in a grid across the entire backplane. However the introduction of *optical planes* vertically aligned with the cache pin-outs made that design obsolete. However the general concept of contacts covering the backplane surface can be resurrected if those contacts are used for power management rather than data transfer.

At the same time a new feature of the cube is introduced as a logical outgrowth of the general design aesthetic. The cube is removable. It can be picked up and handled as an art object. This feature allows the uninterrupted display of the voxel image even after the cube is removed from the backplane. The user can closely examine the three-dimensional volumetric image by turning the cube any way he or she wishes. Only one face of the cube is opaque, which is the base plane.

This removable feature requires a battery backup system that resides in the cube itself. If the cube's base plane is extended from one functional layer to three (see the diagram below), then the batteries and power support can be dedicated to one of those layers.

The power requirement for millions of micro-LEDs isn't small. Given a power consumption of 50 μW for each micro-LED, then there are 20,000 pixels per watt. Therefore,

128^3 / 20,000 = 104.8 watts, all LEDs on, one color only

Luckily, not all of the LEDs will be on at the same time. But even if they are, variable resistance will reduce the load when certain color values are dimmed. There are three micro-LEDs per voxel, for red, green, and blue. Not all will be operating at full power. If a 66 % brightness is assumed for each triple, then an average rating for all three would be:

50 μW x 3 (RGB) = 150 μW x 2/3 = 100 μW, per RGB triple

The nominal LED power consumption for a voxel image is therefore 104.8 x 2 = 209.6 watts. This power rating is far higher than that of a mobile device like an iPhone, but there a few things going for this scheme. First, the batteries will be recharged each time the display cube is placed on the backplane. Second, the length of time that a cube will be handled for close examination is just a minute or two. Therefore a charge should be able to drive this big load for a few minutes if the batteries are designed properly. An auto-shutoff will prevent the batteries from being drained.

The optical receiver circuitry, estimated to consume about 5 watts, is disabled when the cube is lifted from the backplane and goes into *powered separation mode* (PSM). Battery power preserves the FIFO memory on the base plane that feeds all of the color planes. The micro-LED latch circuitry is implemented with nano-components which are extremely power efficient (see Chapter 7). Thus the only component with a big power drain is the micro-LED itself, and micro-LED power-efficiency in the coming years is bound to improve.

A general diagram of the cube's base plane is shown below.

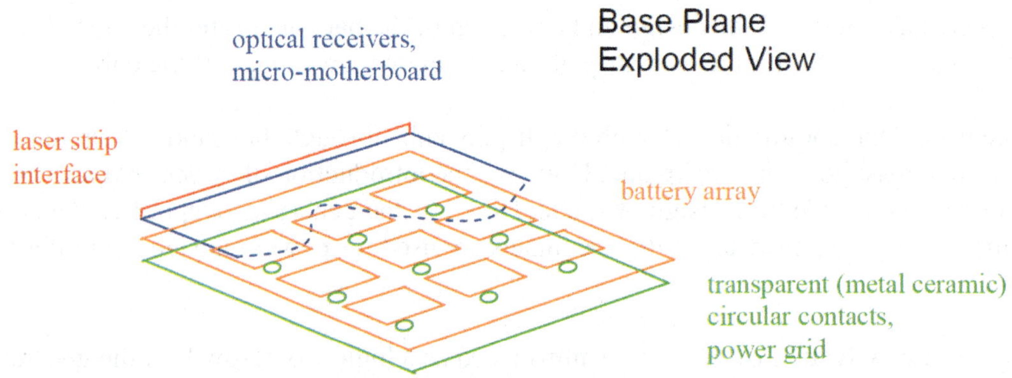

8.2. Next Generation Battery Technology
The most promising and highest capacity battery technology of the future is Lithium-air. But before getting to that, let's start with some less desirable options.

<u>A. Graphene in Copper Chloride</u>
Xu *et al* [1] took a graphene sheet on an SiO2 / Si substrate and attached to either end a metal contact: a gold electrode on one side, and a silver electrode on the other. The whole thing was submersed in a copper chloride solution. Copper cations moving through the solution from ambient heat collided with the graphene to produce electrons. The released electrons traveled across the surface of the graphene to the electrodes, resulting in 450 millivolts measured on an open circuit. After 30 days the voltage dropped by a factor of ten to 40 millivolts. The graphene had decayed somewhat, but overall the experiment was considered a success.

The researchers also hooked six copper chloride cells together in series to generate enough power to light an LED. This is shown below.

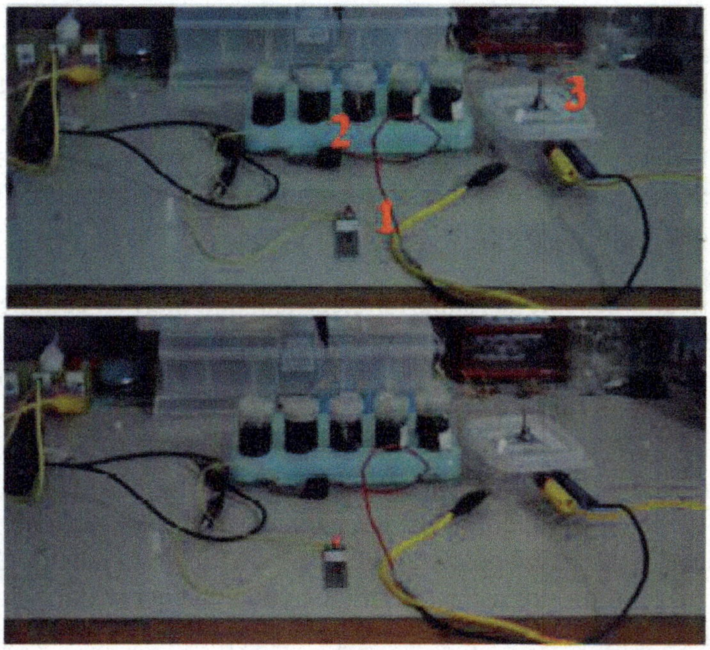

(courtesy [1])

Label 1 marks the LED which is lit up in the bottom picture. Xu *et al* believe this isn't just a chemical reaction, but is truly a product of the motion of atoms across graphene. The drawbacks are: a) the battery system requires a liquid solution; b) the graphene is corrupted after many hours of operation; and c) the sheer volume of copper chloride needed to get a power level suitable for driving a single LED is excessive for something like the cube.

We want to avoid liquid suspensions and keep the cube and the backplane completely solid state and dry. This is a design requirement. The battery package should be well sealed and low-profile, as indicated in the base plane diagram.

<u>B. Ambient Extraction Power Sources</u>
Another class of power sources are devices that extract vibrations and/or heat energy from the ambient environment. For example, a *heel energy harvester* embedded in a shoe consists of a compressible rubber diaphragm that deforms a sheet of electroactive polymer, which creates in turn a voltage spike. This is shown below.

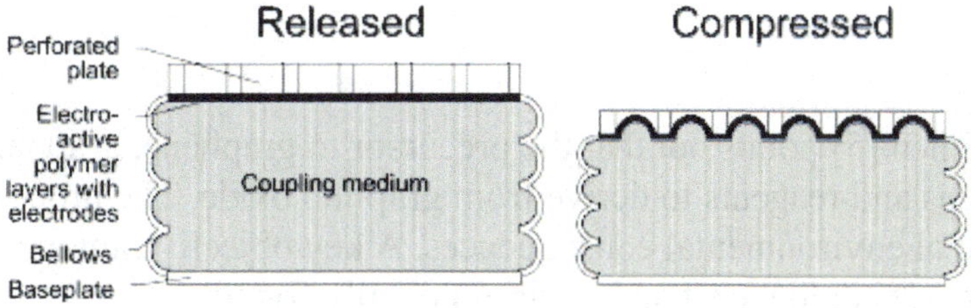

(courtesy [2])

Technically the copper chloride graphene battery is this kind of power source, though the ambient environment (copper chloride) is artificial. The big question for the display cube would be: where is this compressive or vibrational energy coming from? Throwing the cube around is clearly not an option.

C. Rechargeable Lithium-Ion Batteries

Lithium-ion batteries (LIBs) are well known because they're used in the current generation of mobile devices. The power capability of a LIB depends critically on the speed at which Li+ ions and electrons migrate through the electrolyte and bulk electrode. Researchers are trying to: a) improve the electron and Li+ ion transport at the electrode surface or in its bulk; and b) reduce the path length over which the electrons and Li+ ions have to move by using nano-materials.

One such nano-material is graphene. It has superior electrical and thermal conductivity, excellent mechanical flexibility, large specific surface area, and an open and flexible porous structure. Chemically doped graphene used as an electrode material exhibits rapid surface Li+ absorption, and ultrafast Li+ diffusion into the bulk of the material. Two such chemical doping agents are nitrogen and boron. [4]

Battery performance can also be enhanced with graphene composites based on elements like iron, sulphur and vanadium [5]. As cathode materials for LIBs, composites offer high capacity and good cycling (charge and discharge) performance. Typically when you discharge a battery at a fast rate, some amount of the battery isn't used. Graphene composites use more of the available energy during a fast discharge. This is related to the high specific capacity of composites. In general, power is a measure of the *rate* at which the work in the battery is done.

The material below is a 3D iron-graphene composite.

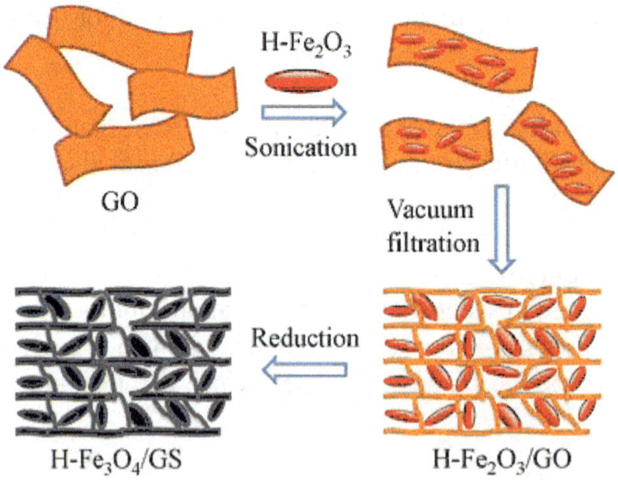

(courtesy [6])

(GS = graphene sheet.) Chen [5] points out that the precursor to graphene, graphene oxide (GO) as shown above, requires many steps and reagents to derive from graphite oxide. The process requires many chemicals and has potential environmental consequences. A key objective among researchers is to make this kind of graphene processing low cost and environmentally friendly.

D. Lithium-air Batteries

This brings us to the most impressive of the next-generation technologies, which is a variant of the lithium-ion battery called *lithium-air* or *lithium-oxygen*. In the lithium-air battery lithium metal acts as the anode, and oxygen freely absorbed from the environment acts as the cathode. Pure oxygen is denser than oxygen in the ambient air and augments the reaction.

An "air electrode" with micrometer sized porosity for oxygen diffusion, and nanometer sized porosity to catalyze Li-O2 reactions was designed by Xiao *et al* [7] using functionalized graphene sheets (FGSs) with well-understood (and desirable) lattice defects. A prototype tested with pure oxygen with no external load showed a huge capacity of 15,000 mAh / g, pronounced "milli-amp hours per gram," at a current density of 0.1 mA / cm^2. This capacity is ten times higher than all the other next-generation lithium-ion designs. The same prototype in ambient air had a capacity of 5000 mAh / g, still impressive.

The problem with lithium-air batteries is that the reaction produces a byproduct called Li2O2 which clogs the air electrode during the discharge and degrades performance. Xiao *et al* produced a unique 3D structure from a colloidal process, that, combined with the graphene lattice attributes, causes the Li2O2 particles to be nanometer sized and isolated. The structure is unfriendly to large particle formation, and so the problem of air channels being blocked by Li2O2 particles is mitigated.

The exciting aspect of this work is that there may be a preferred structure whereby discharge products are minimized. Xiao *et al* produced functional groups (hydroxyls, epoxies, and carboxyls) in the graphene by expanding and reducing graphite oxide under high temperature. The resulting graphene sheet contains certain imperfections called 5-8-5 defects, which in turn discourage the clustering and migration of Li2O2 particles.

At the same time, the unique cellular structure – arrived at with natural surfactants producing foam, i.e. by a colloidal microemulsion process – approximates a yet to be determined regular structure that can bring out the highest capacities of the material. In other words, Xiao *et al* are on the threshold of great nano-engineering discoveries.

The micrograph below shows the porous structure of the experimental material.

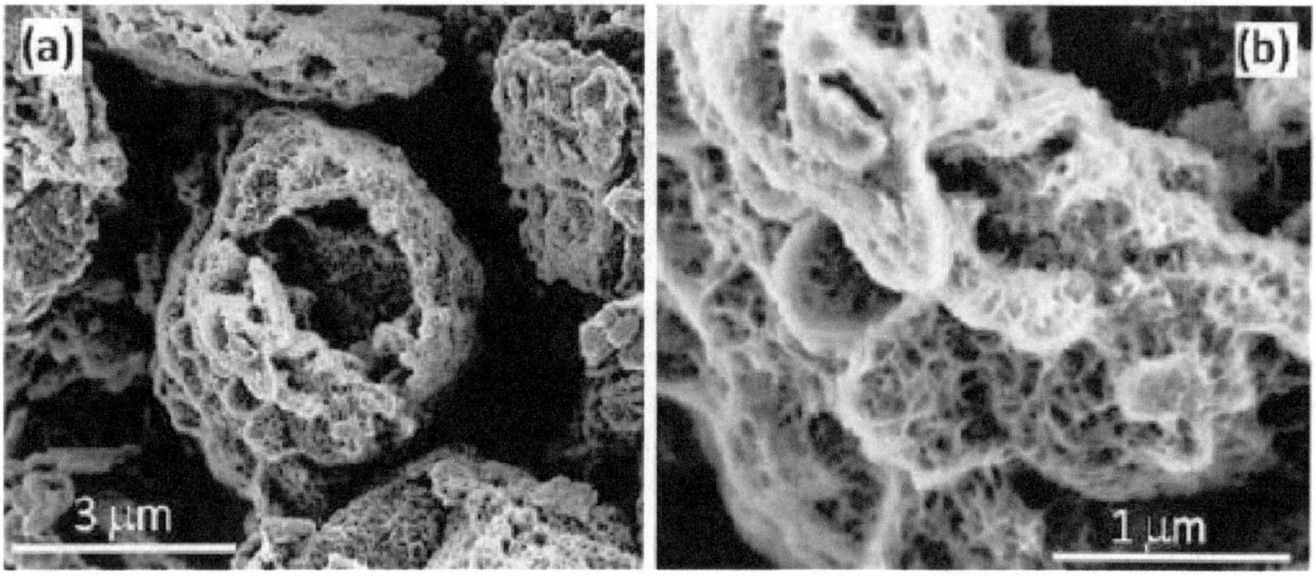

(courtesy [7])

One can imagine how oxygen can freely penetrate this cellular structure, both at the macro and micro scales. Xiao *et al* note that during discharge the large tunnels function as highways to supply the oxygen to the interior parts of the air electrode; while the small pores on the walls are the exits which provide triphase (solid – liquid – gas) regions required for oxygen reduction.

Understanding the roles of the lattice defects and the functional groups (i.e. the binding of COOH) on the graphene substrate is very important because it may be possible to tune the energy / power ratio for specific energy applications through control of the defects and functional groups. For example a battery could be fine-tuned for high charge and discharge rates. Controlling the architecture and surface chemistry of the carbon electrode requires, however, a fuller understanding of the various Li-O2 interactions taking place.

Li *et al* [8] developed a graphene nanosheet (GNS) and used it as an active material in a lithium-air battery cathode. Like Xiao, Li obtained spectacular results, but like Xiao, the Li team observed discharge byproducts in the graphene. Li *et al* paraphrase Xiao *et al* in noting that carbon based electrodes during discharge produce Li2O2 and Li2CO3 because the intermediate product of the reaction between Li and O2 may in turn react with the carbonate solvent in the ethylene-carbonate (EC) based electrolyte.

The graph below compares the GNS with two other cathode materials.

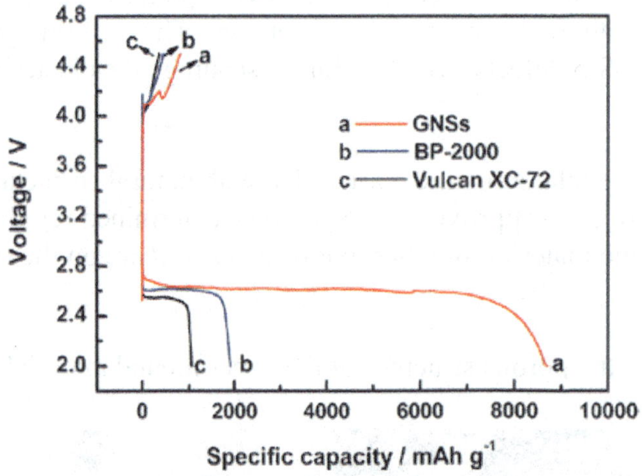

(courtesy [8])

The GNS delivered a discharge capacity of 8705 mAh / g! One reason GNS achieved a better rating than BP-2000 is because GNS has a wide distribution of pore sizes, and a greater number of pores in the 2 to 20 nm size range, whereas BP-2000 has a greater number of pores in the 20 to 50 nm range. Vulcan XC-72, by contrast, has a small number of pores and thus performed the worst.

8.3. Conclusion

We have described another feature of the display cube: the preservation of a static voxel image even when the display cube is removed from the backplane. This allows a user to closely inspect an image while handling the cube.

To pull this off a battery system is incorporated into the basement levels of the cube's base plane. We have looked at some next-generation battery technology, the best of which is based on lithium metal and graphene and graphene composites. We mention in passing the imaging of lithium metal byproducts by Sandia Labs. [9]

The lithium-oxygen or lithium-air battery has the most potential, or if not, at least is the most illuminating in terms of understanding the high capacities that this new generation of batteries produces. The applications for these batteries range from "mobile devices" and "personal movable tools" to "hybrid electric vehicles" and the power densities during fast discharge can be considerable [4]. In any case, next-generation batteries should be powerful enough, light enough, and stable enough to power the entire voxel micro-LED array contained within the display cube for short periods of time.

References

[1] Xu et al. *Self-Charged Graphene Battery Harvests Electricity from Thermal Energy of the Environment*, arxiv.org/abs/1203.0161, 2012, online
[2] Zylka, Pawel. *Current progress in ambient energy harvesting using piezoelectric materials and electroactive polymers*, white paper, 2010, online
[3] Sterken et al. *Power Extraction from Ambient Vibration*, white paper, 2001, online
[4] Wu et al. *Doped Graphene Sheets As Anode Materials with Superhigh Rate and Large Capacity for Lithium Ion Batteries*, ACS Nano, v. 5 no. 7, 2011, online
[5] Chen et al. *Graphene-Based Composites as Cathode Materials for Lithium Ion Batteries*, Journal of Nanomaterials, 2013, online
[6] Sun et al. *Flexible free-standing hollow Fe3O4/graphene hybrid films for lithium-ion batteries*, abstract, 2013, online
[7] Xiao et al. *Hierarchically Porous Graphene as a Lithium-Air Battery Electrode*, Nano Letters, 2012, online
[8] Li et al. *Superior energy capacity of graphene nanosheets for a nonaqueous lithium-oxygen battery*, Chem. Commun. 47, 2011, online
[9] Jungjohann, Katherine et al. *Cryogenic Laser Ablation Reveals Short-Circuit Mechanism in Lithium Metal Batteries*, ACS Energy Letters, May 2021, online

Chapter 9

The Cube So Far

9.0. Introduction

This chapter serves as: an interim summary of previous chapters; an evaluation of some roadblocks to completion; and a look at some lesser-known cutting-edge technology.

9.1. The Design So Far

The display device consists of a *backplane* which also serves as a base for a transparent display cube. The backplane contains display memory, a communications interface to a computer such as a USB port or an IoT port, and the critical *laser strip* which optically couples to the cube. The cube contains a *base plane* situated below the display area proper, parallel to the backplane base. The simplest design contains 128 optical couplers coincident with the backplane laser strip; 128 *clock recovery stations*; and 128 *column headers / controllers* placed at the foot of each column. Each vertical plane aligned with the optical link along the laser strip is called an *optical plane*.

The general form factor is shown below.

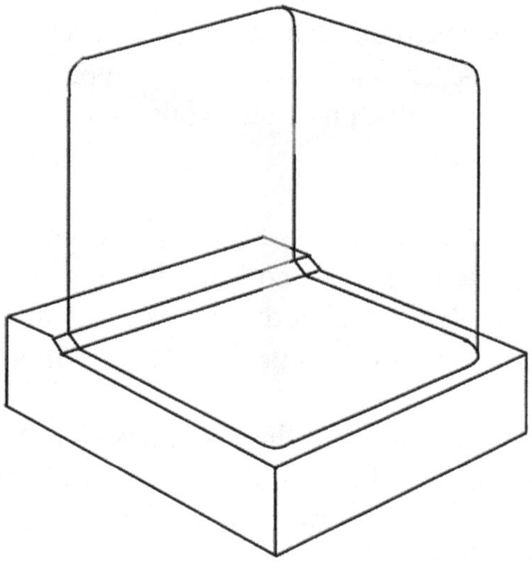

The cube size is nominally given as 2 inches to a side, though this may change when the actual components are finalized. The interior of the cube consists of 128 optical planes, each containing 128 columns. Each column contains 128 vertical levels, each level corresponding to a voxel. Each column receives a data packet with data for all 128 vertical elements.

Each voxel is an active area containing switching and latching circuitry; a digital to analog converter; and a triple micro-LED – one LED per red, green, and blue element. The base plane memory is designed to retain color data when the cube is removed from the backplane, to facilitate close examination of the voxel image. This requires battery backup power in the cube.

9.2. Technical Roadblocks

The journey from the optical coupler to the base station for each column may require signal splits. One technical concern is the reduction in signal power with PCB splits. The effect is illustrated below.

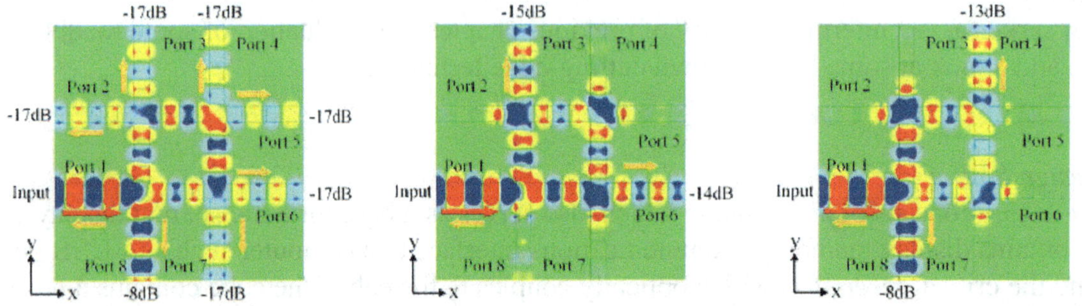

(courtesy [1])

Though this simulation was made for plasmon-scale switching, which is too weak and localized for the distances of the cube, the principles of power loss are the same. Notice how the power gain decreases at each junction of this 8-port network.

The display cube employs two cascades: a high-speed cascade in the base plane to capture the burst bytes for a selected column and its 128 levels; and the low-speed cascade in the column to latch through the level data from each column header in turn. The base plane cascade ensures that only one column header at a time will be gated to the high-speed bus. Thus the power loss will only be "times one" for that one gate. All of the other gates will be closed. But in reality, all of those other closed gates draw current and thus dissipate the signal by their very presence.

Once the signal is conducted up the column, the low-speed cascade ensures that only one voxel at a time gates color data. As well, the low-speed bus is isolated from the high-speed bus. This means that the power loss is again limited to "times one" for that gate, but the same caveat applies here as for the base plane fan-out. Realistically, it may be necessary to boost the signal at intervals in the vertical direction.

Less a roadblock and more a challenge are some of the manufacturing details. One such challenge is the layering registration and fusion process to make a solid cube. Is boron glass really the right substance for the cube? Glass can be made with a high optical transparency; and yes it's an insulator. However how will the glass be fused without fusing the electronics?

Another challenge is cooling. This is related to the layering because the glass will have the same quality as *diamond-like carbon* or DLC, which acts as a heat sink. Imagine those millions of micro-LEDs pumping away and the heat that will be generated. The display cube may get quite warm after some minutes of operation! Along these lines, metamaterials can also generate heat, so again glass must come to the rescue to dissipate that heat.

9.3. Near and Far Fields of the Invisibility Cloak

We now return to the hurdle of invisibility using metamaterials. Progress has been made in plasmon resonance in the near field. *Near field* refers to field effects in the very restricted range of nanometers. The cloak hides transmission lines which run through its center. A cloak works by essentially bending light.

This implies that light can leap across a solid conductor interface at the inner boundary of the cloak. In fact, this has been achieved, both at near-infrared and optical wavelengths, for broad bands and for a wide range of impinging incidences. The key enabling structure is a nano-array: examples include a nano-array of gold-plated spheres, a sequence of corrugated strips, or even a grid of rectangular gold-plated pedestals. [2, 3, 4]

The graphic below comes from the prodigious team from the Jiangxi Normal University and Key Laboratory in China. Their work is not freely available; only one paper could be found online without cost [2]. Their research offers variations on the theme of nano-arrays. The most effective array configuration seems to be hexagonal – whether a hexagonal array of spheres, rectangular holes, or gold elliptical nano-wires. The graphic below shows a hexagonal nano-sphere array. The spheres are made of polystyrene plated with gold.

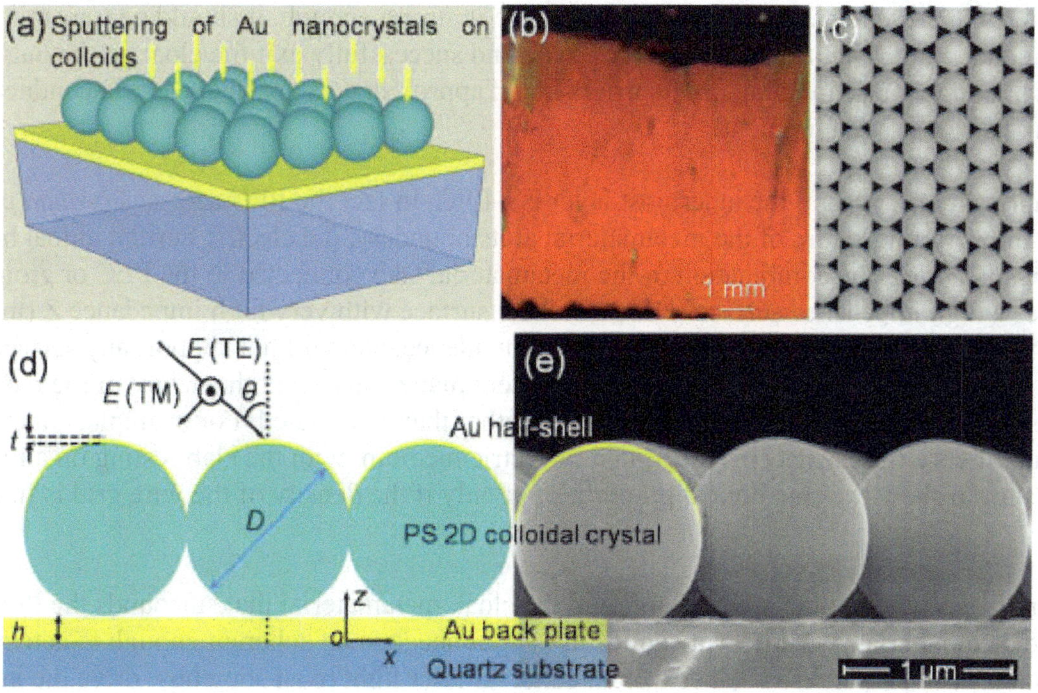

(courtesy [2])

If the thickness of the gold-film backplate is greater than some threshold value (say 100 nm), the sheet acts as a *super absorber* for a suprisingly wide bandwidth. If the thickness is less than the threshold value, radiation is passed through the seemingly solid barrier due to plasmon resonance. Miraculous but true: solid matter does not act as a barrier. The experiment in [2] works in the near-infrared, but the Jiangxi team has simulated configurations in the optical regime with perfect (near-unity) optical transparency. The results of the experimental setup in [2] are shown below.

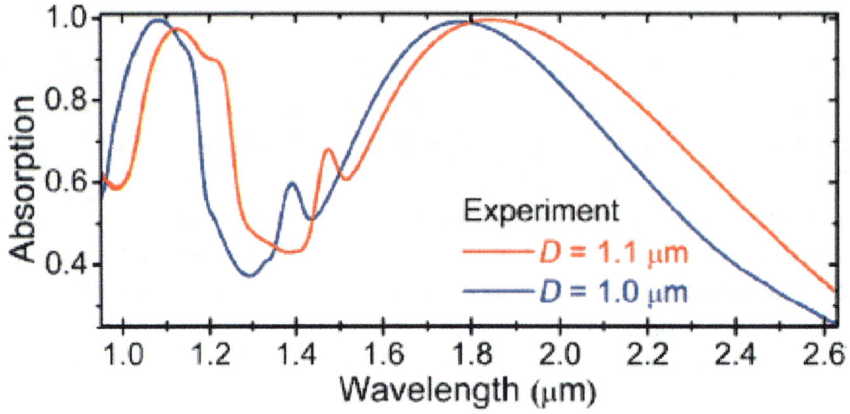

(courtesy [2])

An important insight is that as the diameter of the micro-spheres becomes smaller, the absorption bandwidth is blue-shifted. This suggests that a micro-array that works in the infrared with a given sphere diameter can be redesigned for optical frequencies by making the spheres smaller and the array finer.

The idea is to place this kind of light-collecting array at the interior boundary of the cloak to transmit light according to some clever cloak design. If the resonance is strong enough, it should be transmitted through a continuous conducting layer, travel around the cloak, and successfully exit the cloak to the outside world; and at the same time the conducting gold barrier should approximate a "perfect electric conductor" (PEC) for the cloak (which is a requirement).

In the cylindrical cloak design, the innermost layer is a PEC. In fact the PEC acts as a *ground plane*. Because of the very fine nature of the metamaterial at its boundary, the electric current at that boundary is said to be *microscopic*. The combination of the metamaterial slab connected to the PEC or ground plane characterizes a high impedance surface (HIS). An ideal surface with very high impedance Z (infinity) will fully reflect incident waves in phase for all angles of incidence, and will not support any surface waves. In practice impedance depends on the frequency of incident plane waves and the nature of the metamaterial that borders the PEC. Surface waves are suppressed rather than eliminated. For example, a metamaterial slab made up of very fine wires embedded in a dielectric medium, with the slab resting on a PEC ground plane, will start to show extreme optical properties but only if the density of the wire grid is fine enough. [7, 8]

The problem lies is transmitting the light through the cloak metamaterial that surrounds the PEC. In essence, the metamaterial is a fancy waveguide, and light needs to be coherent enough to activate the special metamaterial properties. In other words, coherent laser light must somehow drive the interaction. This is illustrated with the setup below.

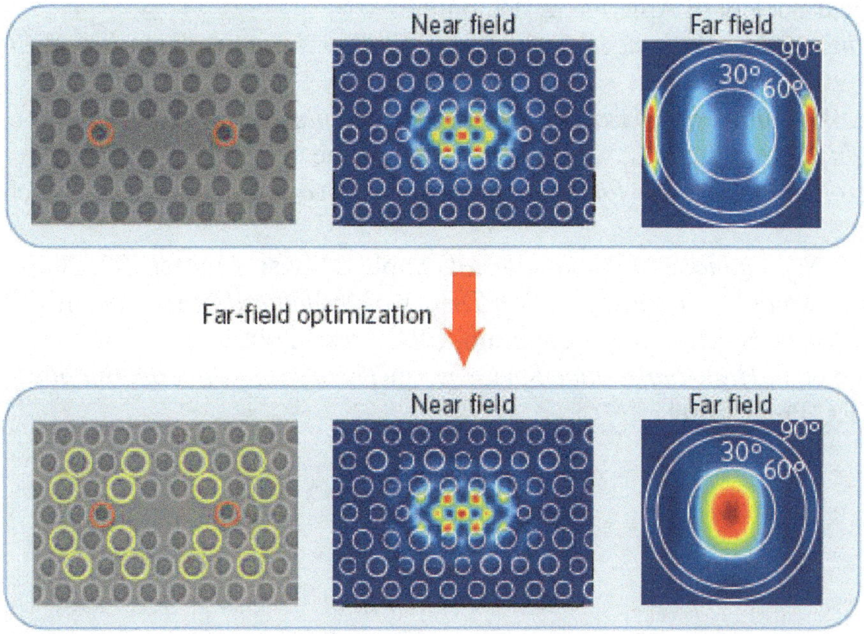

(courtesy [5])

As shown in the top row, a photonic crystal nano-cavity is stimulated with a laser, and the intensity of the light in the far field is scattered towards the edges of a Gaussian distribution. However when a secondary lattice is imposed on the primary, consisting of slightly enlarged holes, the resulting lattice has the effect of focussing the light in the far field to a near-Gaussian distribution.

This kind of nano-cavity design is similar to what a cloak would use. Right now at least, only a laser can stimulate the material with enough energy to move light to the far field.

One solution to this problem is to use superconductors for the metamaterial. Room-temperature superconductors would be best, but the desirable qualities of superconductors can only be achieved, so far, with materials that are super-cooled. For example, the transition temperature of a thin film of yttrium barium copper oxide (YBCO) is somewhere between 85 and 51 degrees K (-306 and -367 degrees F). At this temperature the material becomes *superfluid*. The authors of [6] measured the resonance of a YBCO split ring resonator in a liquid helium cryostat! So much for room-temperature superconducting metamaterials.

9.4. Conclusion

I'd like to thank the readers who made it this far. Also let me acknowledge the hundreds of researchers whose hard work has influenced the design of the 3D display device. The scientific age is alive and well and if the display device is ever built, it will be proof of such. Hopefully this chapter has allowed the reader to catch his or her breath before moving into more detail.

References

[1] Ooi et al. *Mid-infrared Active Graphene Nanoribbon Plasmonic Waveguide Devices*, preprint, 2013, online

[2] Liu et al. *Dual broadband near-infrared perfect absorber based on a hybrid plasmonic-photonic microstructure*, Optical Society of America, 2013, online
[3] Wang et al. *Grating-assisted enhanced optical transmission through a seamless gold film*, Optics Express, v. 22 no. 5, 2014, online
[4] Song and Zhang. *Wide-angle polarization-insensitive transparency of a continuous opaque metal film for near-infrared light*, Optics Express, v. 22 no. 6, 2014, online
[5] Priolo et al. *Silicon nanostructures for photonics and photovoltaics*, Nature Nanotechnology, v. 9, 2014, online
[6] Gu et al. *Terahertz Superconductor Metamaterial*, Applied Physics Letters 97, 2010, online
[7] Silveirinha, Mario. *Nonlocal Homogenization Theory of Structured Materials*, in *Theory and Phenomena of Metamaterials*, Ed. Filippo Capolino, CRC Press, 2009
[8] Goussetis, George et al. *High-Impedance Surfaces*, in *Theory and Phenomena of Metamaterials*, Ed. Filippo Capolino, CRC Press, 2009

Chapter 10

Ambient Illumination

10.0. Introduction

Here we explain how the transparency of the display cube can be used to advantage so that any LED backlight can be done away with. Instead of using an LED backlight, *ambient light* can be used to illuminate the light emitting diode array. This change is minor enough so that most of the design elements for the 3D display device remain unchanged.

The principal advantage of using ambient light instead of a backlight is the reduced power consumption. The main drawback is that when the display cube is viewed in the dark, it will rely on the intrinsic brightness of the LEDs themselves. The cube will most likely be used in naturally-lit office spaces, for entertainment around the house, or outdoors – all cases in which ambient light is abundant. Artificial ambient light will also give a satisfactory viewing experience during non-daylight hours.

There are however some drawbacks. Ambient illumination is associated with color desaturation. The brighter the illumination the greater the color loss. This color loss is mapped with the chromaticity diagram of the *Commission Internationale L'Eclairage*, or CIE. Pure spectral light is mapped to the chart boundaries, while white light is at the center. Colors like brown, which is an orange-red chromaticity at low luminance, are not shown, because chromaticity is differentiated from luminance. And yet one can speak of "relative luminance" as it pertains to the transition from one color in the CIE map to another color within the map. Any two colors in the CIE chart can be mixed via a linear interpolation to produce a line between them: that line represents a range of colors. Likewise, a third color can be combined with various mixtures of the first two colors to produce a gamut of colors within the triangle connecting all three colors which make up the triangle's points. Different display devices have different color gamuts, and those gamuts tend to be drawn away from the boundaries, representing pure colors, because the more light reflecting off of a screen – or through it – the greater the loss of pure color.

10.1. Transparent LCD Display

Transparent LCD display technology is currently in its infancy. The intended uses of such a technology include store windows, retail display cases, kiosks, bus shelters, and classrooms. Store windows can display information on the window pane while a viewer looks through the window at the display behind the glass. A display case can showcase a product and display information about that product on the glass-case enclosure. A bus shelter can be constructed that lets light in for safety and comfort, but the glass walls can at the same time act as information sources. The graphic below is from the Corning glass company and is part of their "Day of Glass" promotional campaign.

Futurists think that corporations and schools will also use transparent LCD display technology to provide interactive information services for employees and students.

The high transparency ratio of these LCD arrays allows a person to look right through the panel. A conventional LCD panel uses a backlight unit. A *transparent* LCD panel takes advantage of ambient light such as sunlight and consumes 90 percent less electricity than a backlit display. For this reason alone, the use of ambient light is well-suited to a three-dimensional layout with millions of voxel elements.

An iPhone or iPad display provides a backlight along with touch interaction. Though touch interaction isn't a part of the 3D display cube, at least one company offers touch interaction as part of their transparent LCD product. In a retail environment a customer can interact with the display to find prices and product options. The following is a promotional mockup for a retail application with touch sensitivity.

The transparent display shows what any conventional display does, including videos and photographs. Marketers claim that high definition picture quality is guaranteed even under bright ambient light conditions. This is probably optimistic and points to the concern that application engineers have about picture clarity. Using nature as a guide, one can observe that quartz crystal infused with iron has a color spectrum that ranges from dark brown to light yellow, and as such is a natural color-graded transparent display. The colors in the crystal have better definition and contrast with increasing light levels. This is because nature has infinite granularity, so there are no planar surfaces within the crystal to reflect light and reduce the color gamut. Unfortunately technology has not caught up with nature.

In reality most transparent LCD displays today have a *haze factor*. Due to the coatings and electronics on the glass, the haze factor can range from less than 2% (highly transparent) to 30% (somewhat cloudy) – although 2% is an optimistic figure. Achieving a clear field will be a challenge for the cube design – but even a quartz crystal with milky transparency conveys all of the information in an image with less loss of definition than might be expected.

In 2014 Samsung started to produce translucent LCD displays in both monochrome and color models. Though the displays are in limited production, the technology will probably work its way into homes and businesses within a decade. Currently the displays aren't completely clear, but they're translucent enough for a viewer to be able to make out objects and people on the other side of the display while there's an image on the screen.

Samsung sees this technology as being useful for shop windows, billboards, showfloors and showcases. The Samsung displays offer 500:1 contrast ratios under the best conditions and run at a native resolution of 1680 x 1050 pixels. Because the displays have no backlights the LCD panels are energy efficient. A display's contrast ratio is a photometric measurement and for a transparent LCD, the best contrast ratio is achieved when mounting the display in a window, with outdoor light coming from behind and with the indoor viewer looking outside. The contrast ratio for the display measured under these conditions in [4] was 138:1 and the related ambient color performance only covered 4.5% of the RGB CIE diagram!

10.2. Graphene for LCDs

A liquid crystal display contains an array of cells. Each cell consists of a thin layer of liquid crystal sandwiched between a pair of polarizers crossed at 90° to each other. Light entering from behind a cell (via a backlight) gets polarized in one direction when it passes through the first polarizer, so that when it reaches the second it can't get through. This makes the cell appear dark. To make the cell bright, the light must pass through the second polarizer. To do this the light must be rotated by the intervening liquid crystal.

To this end, an electric field is applied across the polarizers and this twists the orientation of the long molecules in the liquid crystal. Light rays are guided along the twist of the molecules, and by the time the light reaches the second polarizer it has rotated through 90° so that it can pass.

The electric field has to be applied using electrodes, and these have to be both transparent and good electrical conductors. For these qualities engineers turn to *indium tin oxide* (ITO). However, this material has its drawbacks: indium is rare and therefore expensive; and ITO can release both indium and oxygen ions which prevent the liquid crystal from aligning correctly.

In 2008 a team including Andre Geim and Kostya Novoselov from the University of Manchester in the UK and Sergey Morozov from the Institute for Microelectronics Technology in Chernogolovka in Russia found that graphene is generally more transparent than ITO and has fewer drawbacks.

Graphene composed of a sheet of carbon just one atom thick can be thought of as a single layer of crystal graphite. Indeed, graphene is often fabricated by ripping a layer off of a thin piece of graphite with sticky tape, a process known as micromechanical cleavage (or the *Scotch tape method*). Since Geim and colleagues discovered graphene in 2004, the scientific community has jumped onboard and proposed using it as an electrical and thermal conductor, a semiconductor, and as a sensitive mass detector.

One benefit of using graphene for LCD electrodes is that, unlike ITO, it is stable. This prevents it from releasing ions into the *alignment layer*. The alignment layer is deposited onto LCD electrodes to encourage the liquid-crystal molecules to align properly, but stray ions can reduce the effectiveness of the alignment, causing undesirable *image sticking*. When this happens pixels that should be cleared by electrical alignment remain dark or grey.

More importantly, graphene has better transparency than ITO. Geim's team used micromechanical cleavage to deposit flakes of graphene onto a glass slide, which they put under an optical microscope. They found that graphene has an optical transmission of about 98 percent – significantly higher than the 82 to 85 percent of standard ITO. (98 percent transmission is equivalent to a haze factor of 2 percent.)

What is doubly impressive about graphene is that it can achieve high optical transmission with a corresponding sheet resistance (a 2D measure of resistivity) of just 6 kΩ. Using polyvinyl alcohol in the alignment layer reduces the sheet resistance to 400 Ω. Further chemical doping can reduce the sheet resistance to 50 Ω. ITO, on the other hand, has to trade resistance for transparency. If an ITO electrode is made thin enough to rival the transparency of graphene, its sheet resistance skyrockets.

The one downside of graphene is that, in the past, it has been difficult to produce. Micromechanical cleavage can only produce a few flakes at a time and is therefore unlikely to ever be employed commercially. However, Geim's team has developed a new technique that holds promise for producing graphene sheets in larger quantities.

They begin by placing crystals of graphite in a bath of *dimethylformamide* (DMF) and then sonicate it with ultrasound for over three hours. Graphite is hydrophobic which means it tends to clump together in water, but in DMF the sonication causes it to dissolve into flakes. Next, the mixture is centrifuged for 10 minutes to remove thick flakes. The mixture is subsequently sprayed onto a glass slide to form a graphene monolayer. Finally, the slides are annealed for two hours at 250° C in the presence of hydrogen and argon gas. Although the thickness is not consistent over the slide – it varies between one and four layers of graphene – the optical properties match those of graphene produced by micromechanical cleavage.

The graphic below is an artist's conception of a graphene sheet as an ideal monolayer. In reality, the consistency of the sheet using current production methods would be less wavy and more splotched.

(Credit: Jannik Meyer)

10.3. LCD Transparency, A Tutorial

A regular transparent LCD consists of an array of pixels and subpixels with the following typical layout:

Each red, green, and blue sub-pixel is bisected by a clear area, shown in white, that admits full-spectrum ambient light. In the image above, the green sub-pixels are turned off, so that the displayed color is magenta. Notice that when a pixel is off the clear window is also made opaque. One presumes that any non-zero value of red, green, or blue turns the transparency window fully on for that sub-pixel. In this way ambient light is used to illuminate the liquid crystal array. However, since white light is a mixture of red, green, and blue light, then placing the transparent window next to, for example, a red sub-pixel, has the side effect of mixing transmitted white light with the red, resulting in the red sub-pixel showing up as a combination of red with small amounts of green and blue. This spillover also affects blue and green sub-pixels.

The problem is that the external illumination impacts the color gamut of the display. The red, green, and blue primaries tend to look washed out. Penczek *et al* [4] have come up with an ambient lighting model for a transparent liquid crystal display that takes into account light reflected off of the front of the display, and light transmitted through the back. Penczek found that under a variety of lighting conditions the transparent

LCD performed worse than a normal tablet display – mostly due to the light transmission through the back. The color-gamut areas for the transparent display were much smaller than for the tablet displays. The authors explain, "The clear mid-section of this subpixel design washes out the transmitted colors by allowing broadband light through the color-filter array and severely limits the display's color gamut."

10.4. A 3D Approach

Penczek's work only deals with light reflecting off the display surface, and light being transmitted through the transparent subpixels. The Penczek illumination model is two-dimensional. A 3D display cube would complexify the equation because three dimensions are involved. As well, the viewing angle can change as the user handles the cube.

However the basic idea of a transparent window within each sub-pixel to provide ambient illumination is important. From both a design and a manufacturing viewpoint, the following design for a 3D transparent system seems workable. Each voxel in the 3D system is described by three rings: one each for red, green, and blue. The rings lie horizontally – that is, they are parallel to the base plane of the display cube. In between each ring is a spacer plane that admits ambient light.

The cube is partitioned into 128 vertically-oriented optical planes. Each plane consists of 128 columns aligned with 128 gated *base stations*. This design still holds in the abstract sense, but now the design can be made more concrete. Each optical plane is only conceptual – the physical reality is 384 *ring planes* that are oriented horizontally. The physical layer is in fact horizontal, not vertical. The cube is divided into red planes, green planes, and blue planes; and 384 spacer planes that separate each ring plane for ambient light transmission.

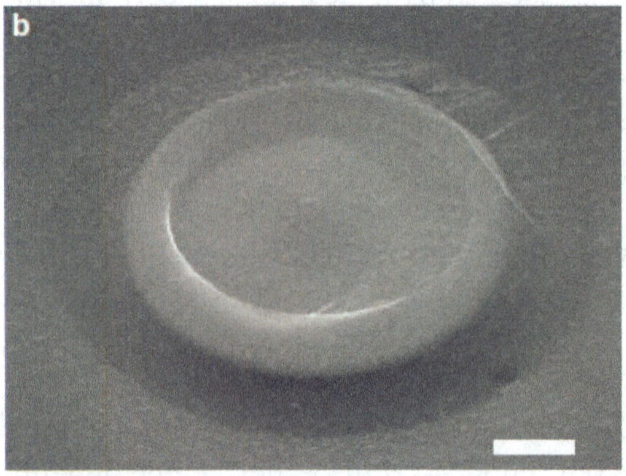

A graphene-covered laser ring: scalebar is 1 micrometer. The application is different, but the idea is the same as for an LCD color ring. (Courtesy [5])

The columns now go from being conceptual to physical: each ring has a processing island, e.g., the color-value data latch and display-element driver circuitry. The cube still uses LEDs, not LCDs, though the driver circuits are nearly identical, so the difference isn't major except for power consumption, but even that can be narrowed down with the latest LED technology. Each data island alternates with a spacer island, making each column a *nano-wafer stack*. Note that the driving circuits reside outside of the invisibility cloak, but if the circuits are small enough they should be close to transparent anyway.

Each ring is a graphene / crystal combination that has activation lines extending from the central island. Each column is essentially a bus, with five signal lines running the length of the column, but it would not be manufactured as a column 128 voxels high. Rather, the column is built up layer by layer as a stack of color planes or wafers. In a later chapter it will be shown that it's not necessary to mold or laser-etch each wafer with 128 x 128 holes. One column can serve 4 rings on each color plane. Conceptually though, each island connects to a red, green, or blue color ring, and the circuitry is identical from one vertical level to the next. The similarity of the ring plane to a silicon wafer should not be overlooked from the tooling standpoint. The graphic below shows the general layout.

Since each control island is surrounded by an ambient-illuminated LED, even if the control circuitry isn't transparent it doesn't matter since looking at the array from outside of the cube, the eye will fuse the rings into colors and the circuitry inside will be obscured. For large empty spaces in the image, the "off" crystal rings should match the near-transparency of the islands and it should all blend into a near-transparent void.

10.5. Conclusion

This chapter has described a transparent display technology that uses ambient light instead of a backlight for image illumination. The ambient approach is a good match for the 3D display device. Note that light-emitting diodes are still estimated to draw about 150 watts for a typical image at a low resolution of 128 voxels cubed. Ambient lighting may or may not help with this power drain. The next chapter will further refine the idea of using color rings.

Graphene is a candidate for the electrodes which make contact with the LED, but it's only a good candidate if the LED stack as a whole is somewhat transparent, and if graphene can be manufactured in quantity and reliably. Ideally, variable resistance will produce varying shades and tints; and just as light is transmitted through natural quartz, so will light be transmitted through the transparent layers of the 3D display cube so that the colors may be seen.

References

[1] intermediatouch.com, Transparent Digital Display Case, online
[2] Corning Glassware, (the various "Day of Glass" promotional videos on YouTube), online
[3] Cartwright, Jon. *Graphene makes for better optical displays,* PhysicsWorld.com, Mar 27, 2008, online
[4] Penczek et al, *A General Framework for Measuring the Optical Performance of Displays under Ambient Illumination*, Society for Information Display, Sept. 2015, online (see also section 11.10 – "Ambient Color" – of *Information Display Measurements Standard*, version 1.03, June 2012, www.icdm-sid.org)
[5] Kim, Yoon-Ho, et al. *Graphene-contact electrically driven microdisk lasers*, Nature Communications, Oct. 2012

Chapter 11
Perception

11.0. Introduction
In this chapter the subject of perception will be explored. When built, the physical layout and configuration of the 3D display device will necessarily dictate a certain look in terms of how the image is perceived. But what is that look?

Luckily, the world of art provides useful information and paradigms for analyzing the possible perception of the 3D display device from a subjective viewpoint. Art offers a visual language for different kinds of images that may or may not be appropriate for a 3D display device. For example, the foreshortened images that we all take for granted in paintings are completely out of place for a three-dimensional volume.

11.1. Getting the Subjective Feel of the Voxel Display
What does the voxel display really look like? This might seem like a dumb question, but it's crucially important. There are many variables in the display cube design, all of which contribute to the clarity and quality of the final image. Even assuming that the clarity and quality of the display are exceptional, "exceptional" doesn't convey the actual feel of the visual experience of looking into a three-dimensional scene.

The closest physical analogue to the display cube is a clear quartz sphere that has orange- and yellow-iron flows and veils in its interior. You can look into this sphere and see a desert landscape, like something on Mars. I've looked at this landscape many times from many angles, and two things are apparent: 1) the work of Mother Nature is very high resolution; and 2) despite this high resolution, the sphere's size (about 2.5 inches in diameter) hinders the three-dimensional definition of the scene. In other words, the smaller the viewing space, the less marked the perception of changing perspectives. The parallax effect of stereoscopic vision is diminished when the viewing extents are less than the span between the eyes' pupils. The display cube will have this same problem because it's about the same size as the sphere.

But worse, the display cube will present a much lower-resolution display than Mother Nature. This will also have an effect on perception. This pixellation effect is a major consideration when exploring visual perception, and is a subject that has already been well-analyzed not by a prestigious computer graphics lab of the twenty-first century, but by Impressionist painters of the nineteenth century.

11.2. Art and the Display Cube
Artists have always been concerned with realism, idealism, and other *isms* (surrealism), but in the

nineteenth century science was becoming dominant and, as author A.M. Vogt puts it, "optics was in the air." [1] For example, Renoir was influenced by color laws laid down in Michel-Eugene Chevreul's *On Colors and Their Applications to the Industrial Arts*, published in 1864. [2] In general the Impressionists were concerned with how color was perceived by the human eye.

According to Vogt, Impressionist painters believed that "truth" lies somewhere within the visible light radiation spectrum. Their mission was to paint an image with light and color to impress upon the eye the truth of Nature. They treated the eye as an instrument or as an apparatus. The Impressionists broke down mixed colors into their primary tones, much as, today, an LED display is composed of red, green, and blue primary pixel elements. The final color is their combined effect. The Impressionists realized that the eye itself does the mixing.

Different artists handled the separation of colors into their primary components in different ways. The most famous system was Georges Seurat's *pointillism*, though Seurat produced surprisingly few color paintings (he's better known for his black-and-white drawings). Painters such as Renoir, Monet, and Degas were attracted to scenes of color and light – scenes that would show off their impressionistic techniques to full advantage.

Claude Monet, *Gare St. Lazare*, 1877. Here Monet shows off the new industrialism as an explosion of light and color.

A painting is a two-dimensional surface, as is a computer display screen. We are fortunate indeed to be able to see high-resolution digital reproductions of classic works of art in our home and with high fidelity. In the old days you had to go to a museum for that, and the experience was sometimes marred by crowds and bad lighting. Today a fine display monitor brings a wonderful saturated backlit quality to digitized artworks.

A three-dimensional volume isn't a good medium for the display of a two-dimensional canvas: the canvas will show up as a thin plane that divides the space in two. In the painting below, Renoir's wife Aline is painted in sharp detail: she seems to pop out from the blurry and impressionistic background. It looks like she's sitting in front of a painted backdrop.

Auguste Renoir, *By the Seashore*, 1883.
A sharp contrast between foreground subject and out-of-focus background.

This painting would be well suited to the 3D display device, with the figure and chair modeled in three dimensions, and with a backdrop as a thin plane in two dimensions behind the chair. But this conjures up a problem: is the backdrop translucent, as voxels lit by passive ambient light would make it; or is it an opaque surface, like a painted canvas or wall?

Consider the painting below of a girl reading. [3] How would this scene be displayed in the 3D cube?

Auguste Renoir, *Girl Reading*, 1876. How would the girl's brightly-lit face be modeled with passive ambient voxels?

Here the light streams from behind and from the left into an otherwise dark room. The display cube, being saturated by ambient light, can't reproduce this scene. Imagine a stream of light in a darkened room: the cube's voxels would have to be perfectly transparent for the light-beam effect to work. The lightbeam would naturally be illuminated relative to the room, but also remain transparent so as not to block the far wall. This is a problem of transparency versus opacity. It requires that voxels in the "on" state be both transparent (the light beam) and opaque (the dark walls). This duality is something the cube lacks. As conceived, the cube is a kind of hybrid: the voxels are partly illuminated by ambient light and are only partly opaque. The bigger problem is that when a voxel is "on" it lacks a black value.

Renoir was a master of color, but he also knew how to paint volumes. The color in Renoir's delightful paintings can sometimes detract from the volumes being painted. Below is a black-and-white reproduction of a painting of Renoir's son Jean. Here you can really appreciate the mastery of Renoir's volume rendering. This subject would make a wonderful volumetric model.

Auguste Renoir. *Jean Drawing*, 1901. Here is an almost photographic treatment of the human figure, but notice that the style is more realism than impressionism.

The picture of Jean Renoir is an ideal towards which to strive. If the display device can show Jean with the same fidelity as Renoir's painting, it will be a victory for the technology. As an ideal, the painting of Jean is useful. But it gets away from the real problem, which is fusing voxel elements together into an image. The display will be an array of individual voxel elements. There's no getting away from this reality. So now we must leave Renoir behind and move into the world of Seurat.

11.3. Seurat, Pointillism, and Pixelism

Seurat drew on pebbly thick paper and the oil pastels, conte crayons, or charcoals didn't fill in all the pebbly holes unless he pressed hard on the stick or pencil. This gives his drawings a pixelated quality that is reminiscent of a low-resolution computer display. Seurat was a master of "pure impressionism" because even though his figures are not necessarily finished, there's no mistaking what's being drawn. Only a master artist can pull that off.

Georges Seurat, *A Music Hall Artist*, 1887-88.

In *A Music Hall Artist* there's no mistaking the singer, the stage, and the conductor in the orchestra pit. This fidelity to the subject matter even under sketchy circumstances gives one hope that a low-resolution volumetric display can convey useful information. The brain is remarkably good at filling in missing visual data.

The problem of light and dark in a volumetric display is brought to the fore in the drawing below, where the woman's form is set against a dark background. If everything was reversed – a light background and a mostly neutral figure – then the cube might be able to display something discernible. Suppose that the default "off" state of the cube is a black background – nothing is illuminated. In that case the figure below represents voxels that are "on" to some range of values, and hence "white." However the problem is that a 3D display must be perfectly transparent, which means that the background isn't really dark: it's transparently white! A cube viewed in a dark room will reproduce the drawing below perfectly; but a cube held before a window in daylight will not.

Georges Seurat, *The Seamstress*, 1882. Light is light and dark is dark, or are they?

Seurat worked with conte crayon on white paper, so the untouched paper is the least variegated: it fuses in the eye as uniformly white. Light will also fuse in the eye when it's bright; so that Seurat's drawings mimic the effect of the underlying cube "grid" becoming more perceptible from voxel to voxel as the voxel illumination levels drop below a certain threshold. With the pebbly paper, the application of conte crayon breaks the uniformity and also lends a gridlike appearance to the shades and volumes.

Georges Seurat, *Woman Reading by Lamplight*, 1881-83. Seurat's crayon creates texture; the paper is uniform when untouched by the artist.

Looking at the cube in natural light, the scenes where most of the elements are "off" will seem the most luminous, because there's nothing to intervene and interfere with the essential elements. Scenes where most of the voxels are illuminated will be hazy and indistinct because "on" voxels partly block each other. Even though a block of "off" voxels should be transparent, in reality the micro-circuitry isn't completely transparent. Differences in brightness, as opposed to color, have the best chance of cutting through the haze, but here again the cube design presents a problem, since the luminosity scale is incomplete, being phase-shifted towards the light-end by ambient light. Light against light doesn't present enough of a contrast to make a volume stand out. But if the design can somehow be improved to provide a full luminance scale, scene contrast and differentiation would be best defined by luminosity differences rather than color differences. As we saw with Renoir's portrait of Jean, the volumes are most convincing when color is absent, because color tends to detract from the essential forms.

Georges Seurat, *Woman Reading in the Studio*, 1887-88.

Here the woman's form might be rendered with "on" voxels, some being of a dark hue; while the light background corresponds to all "off". However if the cube is being viewed in a dark room, everything would be dark. So then the background voxels would have to be illuminated – but illuminated in a way that doesn't block the central figure. In other words, the illumination would also have to be transparent! This is the same problem as with the lightbeam in Renoir's *Girl Reading*.

Below, Seurat frames the woman's head with light on the left, and darkness on the right. The image shows a woman with her head bent over a book. There's enough visual information to discern that she's sitting parallel to the table, with the book in her lap. The light is clearly coming from the right of the canvas. Now, how would the book be illuminated in the display cube? Seurat places its whiteness (plain paper, equivalent to voxels "on") against the black torso of the woman (all "off"). But for the cube, "off" means transparent. In this case, we want "opaque" and "on" and "dark".

Georges Seurat, *Woman Reading*, 1882. This drawing looks photographic from a distance; like something shot on old-style grainy Kodak film (ASA 400).

11.4. Literal Space

Another problem with the cube is that it represents a fixed volume of space. What happens when the "literal space" of the cube ends? The deep space in an image like the one below is cut off at the cube's borders – suggesting as an alternative a compression of literal space into a subjective layering of elements that, as a pseudo-2D canvas, emulates the literal reality. But this layering only works in one direction – the direction that matches the perspectival view into the distance. Rotate the cube even a little and the illusion is broken.

Georges Seurat, *Plowing*, 1882. The cube, being a physical volume, has a hard time with distant horizons, unlike a 2D sheet, where the artist creates an illusion of distance.

For a volumetric display, we're most interested in "literal depth." In the factory image below, the Moon is obviously outside of the cube. Taking the Moon's distance literally, it's outside of the entire Earth!

Georges Seurat, *Courbevoie, Factories by Moonlight*, 1882-83.

Smaller distances are also a problem. Below, the people and factory building are widely separated – too widely separated for the cube, unless the whole scene is scaled down to antsize, and then what's the use? Instead of a work of art we end up with a map.

Georges Seurat, *Men Before a Factory*, 1882-83.

Even something as simple as "a sky" presents problems for the volumetric display. A horizon in the distance, with wagons on a road in front, may work on a canvas, but when turning the cube one would expect the skyline to be replaced by a coast, or by foothills or plains or mountains. Even on a flat Kansas field with no obstructions, turning the cube with a sky at sunset should reveal a darker sky at 90 degrees, and an even more muted sky all the way around at 180 degrees. This means that the original view of the wagons, 180 degrees around, should be illuminated by the sky in the foreground, if the depth of field is great enough.

The cube isn't really suited for the display of scenes with long vistas, which are better suited to a 2D canvas. The best we can do is to emulate the view from a single perspective by foreshortening the geometry in the cube for that one perspective. Turn the cube and the illusion is broken.

11.5. A Design Improvement

Has this study of impressionistic artworks been fruitful? Indeed it has, since certain key deficiencies in the cube display design have been identified. How might the cube design be improved?

1. Can a voxel be electrically illuminated and remain transparent? Transparency will be essential for "lightbeam" effects.

2. Can a voxel be "on" and "opaque"? This contradicts the first item, in that a voxel that's "on" must now be able to be transparent or opaque selectively. Each voxel would have three states:

   a. "off" and "transparent"
   b. "on" and "transparent"

c. "on" and "opaque"

3. We also want a voxel to be "on" and "dark" – with "opaque" implied, as in the dark torso of Seurat's *Woman Reading*. There seem to be two complementary factors at work. One might identify them as *luminance* and *chrominance*. These two factors are at odds with each other. Chrominance can increase in a positive direction or decrease in a negative direction. Remember that chrominance is a function of how the display fuses individual red, green, and blue elements. High chrominance in either direction implies higher opacity. Colors moving towards the primaries, such as Renoir-like colors, will "block" transparency. Likewise, dark hues will block transparency. Both types of color are theoretically "on" and "opaque." In bright light the dark torso of the woman must remain dark. This leads to a new model for voxel states:

a. "negative chrominance" and "opaque"
b. "positive chrominance" and "opaque"
c. "zero chrominance" and "transparent"
d. "high luminance" and "transparent"
e. "positive chrominance" without "primary bias" and "opaque"

Item e) is needed for something like brushed aluminum, which would be a bright reflective surface without chrominance, that nonetheless is opaque. The idea then is chrominance / opaqueness at both ends trading off with luminance / transparency in the middle. Brushed aluminum is a special case that combines high luminance and opacity.

4. Is it possible to automatically adjust luminance based on the current ambient light levels? If the cube is moved into a slightly darker environment luminance should increase. Overall luminance should decrease when moving into a lighter environment. The cube could be outfitted with a photo-coupler to facilitate the continuous adjustment of light parameters across a range of ambient light levels.

5. It may be that all of these problems can be solved with one additional component. Referring to the diagram from Chapter 10, reproduced below with minor additions, opacity may be as simple as making the space between the central control island and the color ring a *sous-bande* (meaning a band below, or like a *sous-chef*, a component secondary to the color ring – the strict English translation would be *sub-band*). It would be a grayscale LCD that, when "on", reflects light but has no intrinsic luminance. However, the grayscale value it reflects is a function of the color value. That is,

$$gsv = (R + G + B) / 3$$

This grayscale value (gsv) is applied to all three sous-bandes, one for each color element, and will range from black (gsv = 0) to white (gsv = max bit resolution). Since the value is a uniform backing value, the collective term *sous-bande* refers to the entire voxel. For example, an (R, G, B) value of all red, (255, 0, 0), would have a sous-bande gray value of (255 / 3) or 85. When "on" the sous-bande would have the effect of blocking light from other voxels. When "off" the sous-bande would be transparent. The list of requirements can therefore be simplified as:

a. "zero chrominance" and "transparent"	[all off]
b. "chrominance" and "transparent"	[color rings only]
c. "chrominance" and "opaque"	[color rings + sous-bande]

Brushed aluminum is now achieved by activating all three color rings (R, G, B) to produce a white color. Additionally, the sous-bande is activated to reflect white. The sous-bande, when "on", reflects the voxel luminance as a function of color value, and is a pure reflector having an intrinsic luminance value.* Below, the *sous-bande* elements are shown with a hatch pattern.

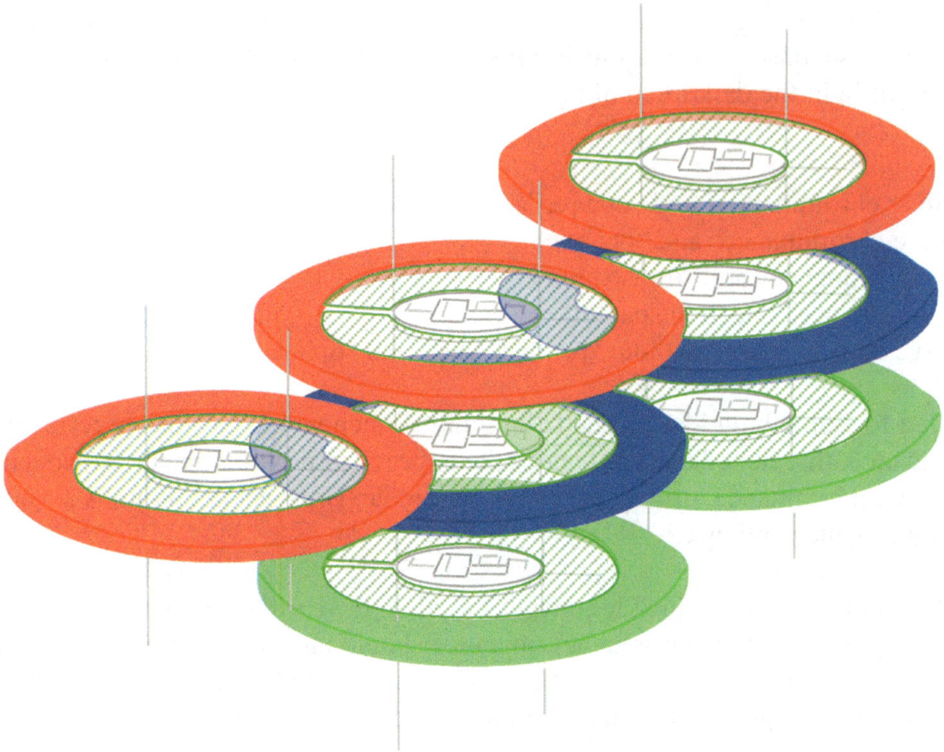

11.6. Conclusion

This chapter has used impressionist artworks to explore how voxel imagery might be perceived. Artistic effects – taken for granted in a 2D artwork – present difficulties for the 3D display cube. The primary difficulties are light/dark transparency and opacity; and views into the distance.

We want to avoid having the display cube become what the 3D printer has become: a tool for making little knicknacks. Go to any electronics store and the front window is littered with 3D-printed gimcracks. The display cube should offer more.

---

* Since the "crystal" in a liquid crystal display is viewable from a whole range of angles in 3D, not just from above as in a 2D display, the *sous-bande* will be a challenge to make. Existing LCD technology uses a magnetic field to rotate the crystal to be opaque or transparent. An alternative approach would employ plasmon resonance in a surface of nanospheres, driven by an appropriate electrical charge to change their absorption and reflection values. Resonance would induce luminance and opacity at the same time.

The analysis of artwork has provided a valuable start towards an enhanced viewing experience. The addition of a single component – a *sous-bande* – addresses most of the display's limitations in this regard although actually making it will be a challenge.

References

[1] Vogt, A.M. *The Nineteenth Century*, The Herbert Press, 1989
[2] Rayfield, Susan. *Pierre-Auguste Renoir*, Harry Abrams, NY, 1998
[3] Hayes, Colin. *Renoir*, Spring Books, London, 1961
[4] Herbert, Robert. *Seurat's Drawings*, Shorewood Publishers, 1962

Chapter 12
Level Design

12.0. Introduction

The previous chapters proposed a nano-wafer stack of central processing islands, with one processing island per color ring. Recall that each ring acts as a color element – red, green, or blue. Fused together as a column, the stack also functions as a vertical bus.

The idea of a stack of processing elements somehow glued together into a column, and the insertion of such a column into each of the ring locations in the horizontally-oriented ring-plane layout, is reminiscent of uranium rods being inserted into a nuclear reactor's graphite-block core. Luckily a bad contact along the column's length won't result in a nuclear meltdown. But realistically, the rods would be so thin that they'd need the tensile strength of titanium to be inserted without breaking.

Another approach is needed. The concept of a "column" need not be abandoned – only slightly abstracted towards a more practical physical solution.

The world is abundant with creative ideas and luckily for the 3D display device design, an idea exists which is feasible for our purposes. *Chip on glass* (COG) technology is used for various purposes in integrated circuit design. One application that will serve as an illustration is the bonding of an LCD driver chip directly to the glass that the LCDs are connected to.

12.1. Chip On Glass Example

Arguably, the chip-on-glass approach increases reliability of the LCD module because there are fewer connections. Since many different conventional modules are being done away with, there are fewer components and so the thickness of the entire LCD construction can be drastically reduced. But like anything that's integrated to fewer components, if a component fails, it may mean replacement of the entire LCD assembly rather than of just the driver chip. However this is a chance that one is willing to take since the construction of a monolithic device with no moving parts, such as the 3D display device, is done in such a way that the integration of parts should increase reliability rather than decrease it.

Reliability will increase with decreasing complexity, and complexity in this case is a product of the number of connections or physical contacts, each representing a single point of failure. The diagram below shows five connections and five points of failure in the traditional LCD and driver setup.

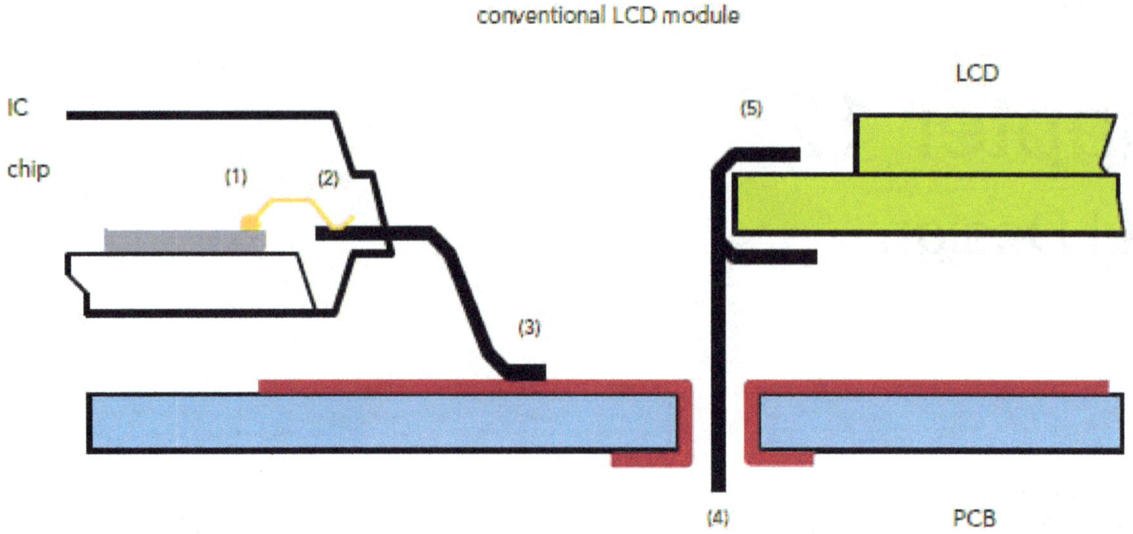

There are five connections between driver chip (1) and LCD glass (5). (Courtesy [1])

Each connection is a potential point of failure. This is what makes the bus in the form of a column so scary for the 3D display device: there are, for a 128 x {R, G, B} x 2 display column (the factor of two is for spacer planes for ambient light penetration) 768 opportunities for failure. Recall that early LCD RGB displays sometimes had manufacturing defects resulting in permanently dark pixels: it's the same idea, but in three dimensions the stakes are raised. There are a cube's worth of failures instead of a plane's worth. So we must think carefully about the best solution.

The diagram below shows the alternative to the conventional approach:

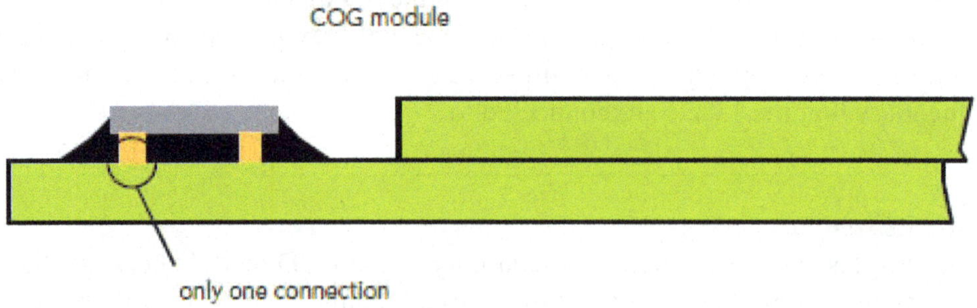

(Courtesy [1])

There is only one set of connections. This is the direction we want to take with the 3D cube. But what's the magic that makes this happen?

12.2. Anisotropic Conductive Film (ACF)

Anisotropic Conductive Film (ACF) is a connecting media that bonds the connector pads of the LCD driver to the glass of the LCD panel. ACF consists of an epoxy (similar to UV curable glue) containing electric conductor beads. The film is anisotropic because the conductor beads only conduct the signal in one direction. The contact between the gold pads of the LCD driver chip and the ITO (indium tin oxide) conductor on the glass is assured in the vertical direction, but the magic is that the signal isn't conducted in the horizontal direction, which would lead to short-circuits.

Curing of the ACF glue establishes the necessary pressure between the gold contacts of the chip and the ITO on the glass to ensure a good electrical connection. Roughly 30 beads per gold "bump" assure a reliable contact. The document provided by automobile electronics manufacturer NXP [1] includes the following details: 1) the resistivity of the connector tracks to the LCD driver must be lower than the resistivity of the ITO connections within a conventional LCD cell. The ITO interface connectors (as opposed to the ITO display tracks) consist of wide tracks with low impedance. The display tracks can be higher impedance as long as the impedance is roughly the same for all rows and columns for ideal "balance".

This setup is shown in detail below:

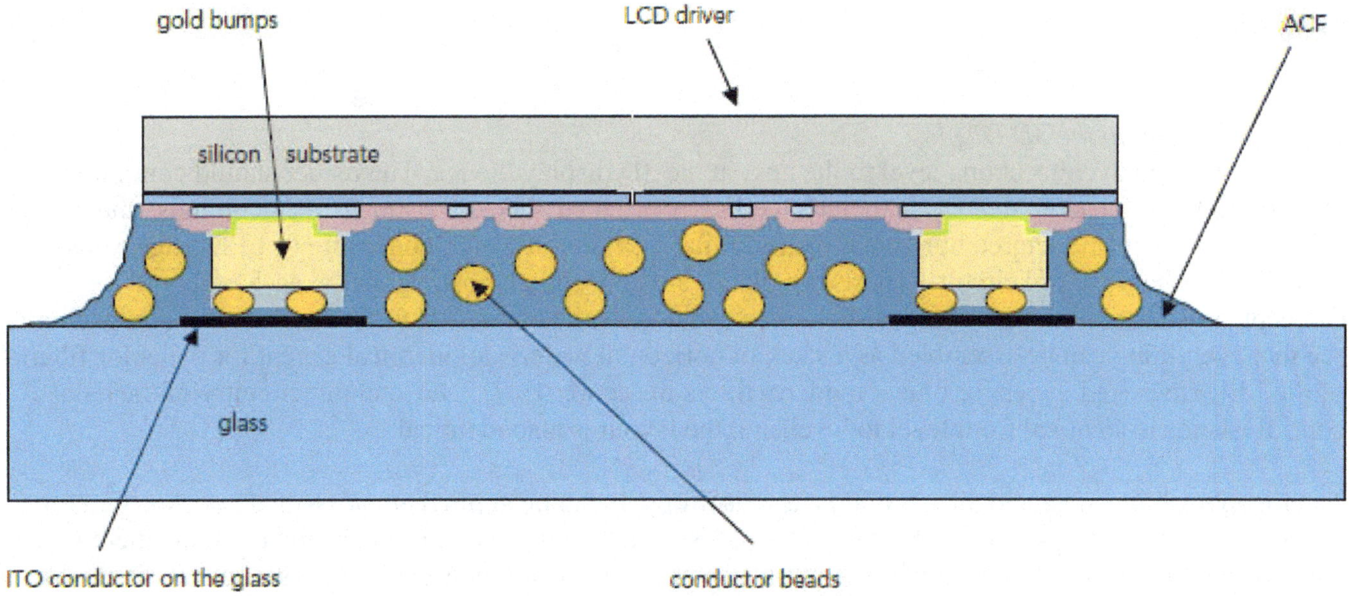

(Courtesy [1])

The glass is a shelf that extends from the LCD itself. There is no horizontal contact between the beads, thereby eliminating short circuits between neighboring gold bumps or contacts of the LCD driver. Apparently the conductor beads are also made of gold. The NXP document allows that "one of the few limitations of COG technology is that only LCD drivers with gold-bump contacts can be used." Similar contacts in the 3D display device will be on the micro-scale or even the nano-scale, which means the economy of scale admits the use of gold, even for a volumetric display.

12.3. Resin Characteristics

This brings us to the characteristics of the resin itself. Recall that the individual horizontal wafer-layers of the 3D display will be bonded together into a transparent mass – irrespective of the ACF used for the column bus. The ACF can't be used for gluing the cube together; only for making contact between bus lines from layer to layer.

However, the ACF must be applied before the cube is heat-treated for melding. This means the ACF must be heat-resistant. As well, the ACF, even in spot applications, must be transparent. There are plenty of transparent glues that already exist so this shouldn't be a problem.

Additionally, the ACF must have a long lifetime. The NXP paper [1] doesn't mention the lifetime of ACF, but from other online sources it's in the ballpark of 10 to 20 years, which for a consumer device like the 3D display cube is acceptable. Lifetime is comparable to existing resin epoxies. As with all plastic compounds, heat will break down the chemical bonds over time.

One other characteristic is critical – reliable contact. There should be enough micro-beads in the formula to form a reliable and long-term contact. The beads shouldn't break down under pressure or heat. In summary, the compound must be:

- Heat resistant
- Transparent
- Long lifetime
- Reliable contact

12.4. Application to the 3D Display

ACF can be used to connect one level to the next in the 3D display device. The reader should continue to distinguish between an *optical plane*, being a vertically oriented slice consisting of 128 columns; and a *ring plane*, being the manufactured horizontal plane similar to a silicon wafer, consisting of LED rings, sous-bandes, and island control circuits. These two planes are conceptually orthogonal to each other. Conceptually, the wafer-stack or signal-column is what links the horizontal to the vertical. But now we will see that everything can be assembled as a stack of horizontal wafers, all identical except for the color filters of the LED rings (red, green, or blue – more on filters in section 12.7). The latching circuitry of the data/control islands is identical from level to level, and the layout is also identical.

The distinguishing characteristic of ACF is its anisotropy. It can be applied *en masse* and not short-circuit because of the anisotropic contact between beads. A simplistic bus design is a hexagonal arrangement of six data/control lines around the data island – but inside of the electroluminescent ring. This scheme is shown below for a single voxel:

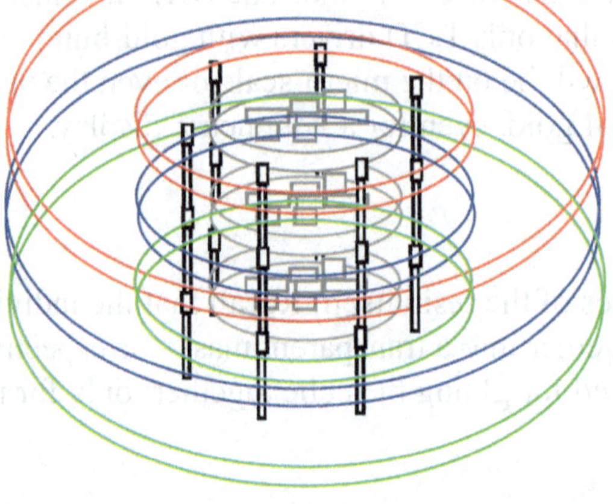

This particular arrangement meets the requirements of transparency, but there are six separate applications of ACF per plane. Each application is a potential point of failure. If the application of adhesive goes wrong, the vertical bus will be broken. Therefore the highest reliability will be achieved with the fewest individual applications of ACF.

The graphene ring and central data island communicate only because the circuitry of the island remains embedded in the wafer. In the vertical direction we want to avoid punching a hole in the graphene sheet that's draped across the island. Nothing should penetrate the sheet. The design shown above, with bus-connectors inside of the graphene ring, will punch through the graphene of the LED. Is there an alternative?

An array of rings naturally has two areas to be exploited: the interior of each ring, and the common area at the nexus of a cluster of four rings. Transparency will be hard enough to achieve with the data island embedded in each ring – so the bus lines should be moved outside of the rings so as not to further cloud the interior area. ACF is applied only once for the 5 x 4 = 20 – 3 = 17 control/power lines that serve all four rings. The power and ground lines can be shared between the four rings. The overall line count is:

1. 4 x lower speed digital serial data bus
2. 4 x lower speed clock signal
3. 4 x IGM/CNB daisy chain signal
4. 1 x power
5. 4 x ground (keep ground lines abundant)

Since for any one group of four rings, two of the rings will be associated with one optical plane, and the other two with the neighboring optical plane, optical planes will share a *base-plane bus station*. This is just the base connector. Even though every optical plane is loaded in parallel, parallelism isn't a requirement for sharing a base station. The signals are gated onto the appropriate lines by the appropriate optical planes at the appropriate times. This is why separate clock signals are used. Only the power line is shared.

The *contact* layout is replicated from level to level, and each level would have an ACF single-point-of-failure *contact cluster* or *quad nexus* as follows:

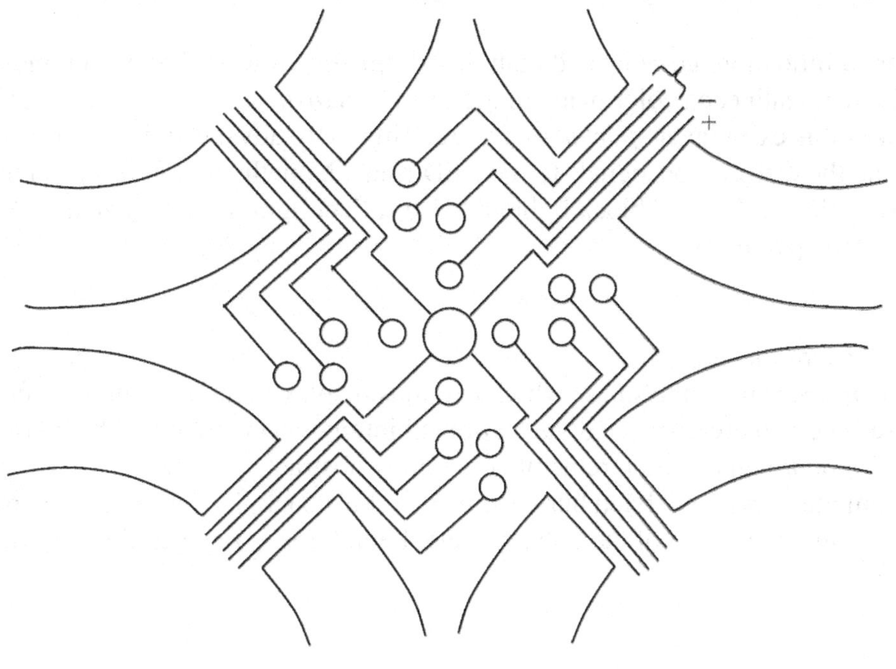

This layout serves, for example, four graphene rings in a red plane. With this nexus design the contact clusters are more sparsely distributed in the ring plane compared to the ring-center approach, leaving empty and thus optically clear areas in the ring layout:

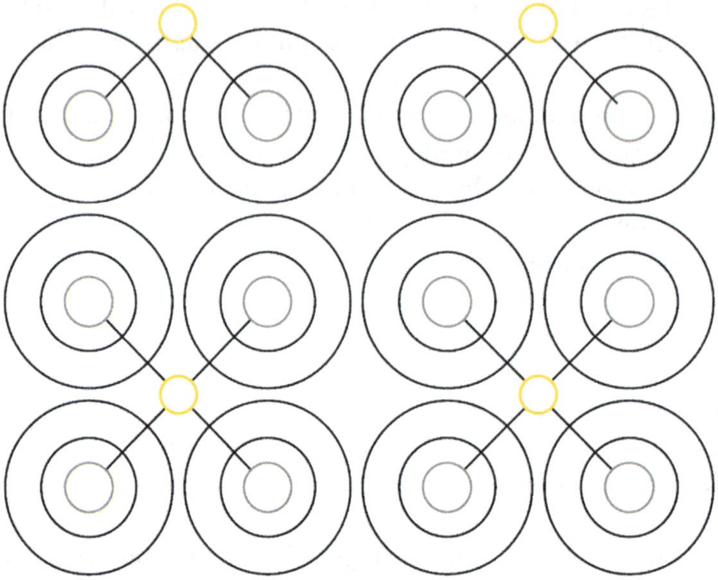

The data islands are shown in grey and the ACF clusters in gold. Each row corresponds to a vertically-oriented *optical plane* projecting out of the page; the spaces between quad groups are nicely empty, thereby reducing clutter and increasing cube transparency.

The lines still need to penetrate the entire wafer, and also will need to be anchored to the wafer to avoid slippage under heat pressure; existing multi-layer silicon wafer techniques can be used to achieve this.

However things are a little more complicated than this simplistic view, in that power and signal lines will have to be routed underneath each color plane in the form of a *distribution layer*. The *cellular layout* of driver circuits makes this extra layer *spartan* or *sparse*. This more advanced design uses through-silicon-vias (TSVs) to route the distribution signals to the LED rings. Nonetheless, ACF is still used for the column design. Though the gold in ACF will block light and also reflect light, the small area of ACF contacts has a minimal impact on transparency.

12.5. From Base Plane to Bus Station

Recall that the *backplane* is the cube's base which contains display memory; while the *base plane* is inside the cube at its base. The two are connected via an optical interconnect embodied by a laser strip. Each vertically-oriented *optical plane*, which remember is purely abstract, is written to in a series of bursts of fiber-optic data from the backplane. Each burst from the cube's backplane through the optical interconnect sends column data to every column of the optical plane. Each burst is recovered via high-speed ECL in the cube's base plane.

Once shifted in at high speed, the data is shifted out at considerably slower speed for each column. We now make an improvement in color resolution by adopting an 8-bit color value instead of 3. Instead of one 9-bit word, with 3-bits each for red, green, and blue, we have three bursts each of 8 bits, giving full 24-bit color. The red values are shifted in first from display memory; then green; then blue: but when scanning out to the slow bus each full red-green-blue value is shifted serially into the data-line, so that the daisy-chain from level to level can be achieved.

The cloverleaf arrangement which defines a voxel nexus requires some basic control logic in the base plane to gate the appropriate voxel data for the appropriate column. There are also a few clock divisions to help send the data at the right time. In the diagram below the labels {A, B} refer to the first and second columns for a nexus for an optical plane. As well, the labels {L, R} correspond to the left and right optical planes connected to a shared vertical bus. Due to the quad layout, the base-plane bus stations alternate on every other pair of optical planes, and every other pair of columns.

High-speed serial-in registers latch the data on the fast clock; and the logic is designed so that those same registers can be clocked out with a second slower clock. These registers are labeled {R, G, B} for red, green, blue. The nice thing about alternating bus stations is that the optical planes naturally divide the base-plane logic into high-speed lanes on the outer margins, and low-speed lanes in the interior were the bus stations reside.

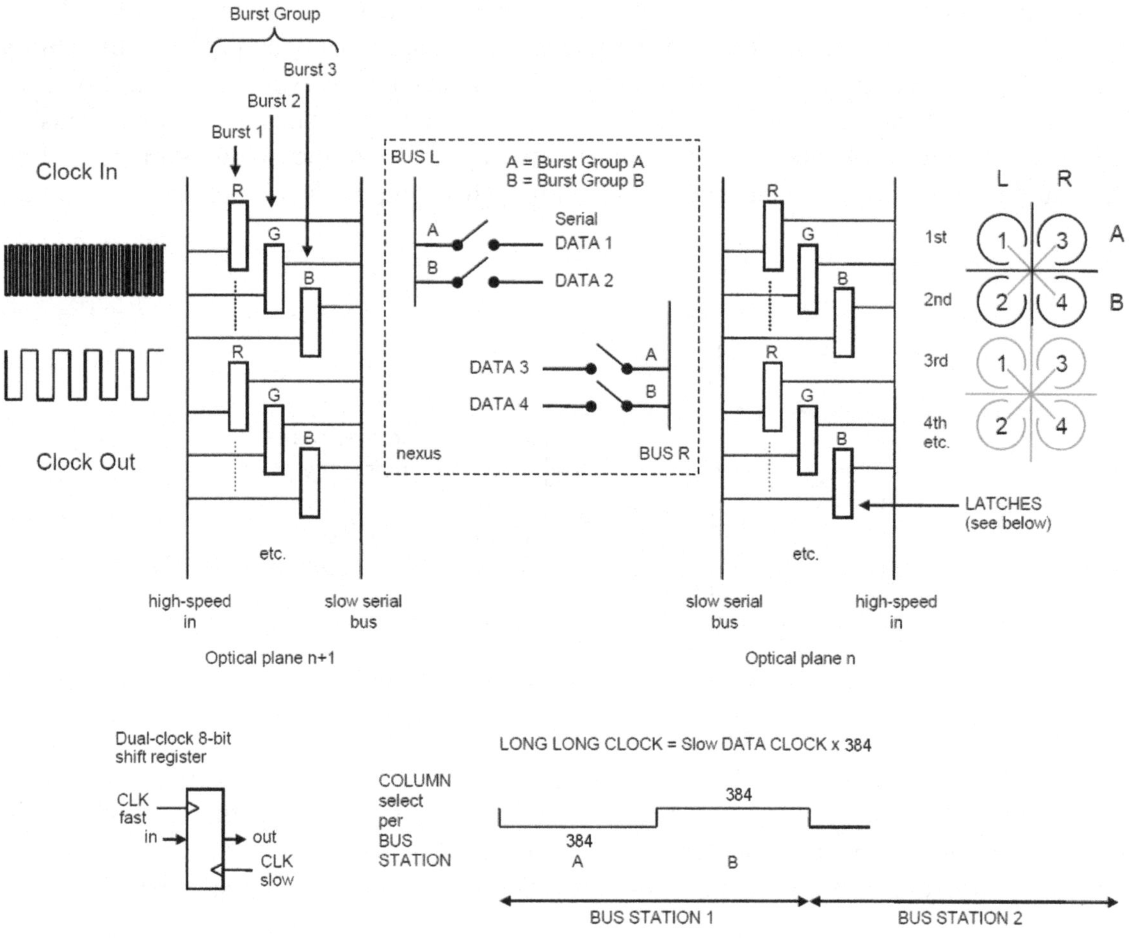

To get through all of the columns of an optical plane, one can think of "finishing" a column in two ways. The in-clock latches 128 values for each burst of red, green, or blue, with each burst filling the optical-plane footprint of 128 *column heads*; and the out-clock has a period 384 times slower to scan out {red, green, blue} triples for successive levels of the current column. In other words, the latency of 128 transfers for each base-plane column provides more than enough time for the slower clock to do its job of emptying the register. Scanning out triples preserves the daisy-chain order of the column levels. This logic in fact would be a little more complicated, but the diagram gives the general idea.

In this simplest design there are 128 / 2 = 64 bus stations per optical-plane pair, and likewise 64 in the orthogonal direction, making 4096 bus or base-plane stations total. This is for the resolution of 128-cubed – the same math applies to other resolutions.

12.6. Spacer Layers and Invisibility

There will need to be some spacing between layers so that the {red, green, blue} elements of each voxel conform to a cubical shape; however these spaces are not air gaps. They consist of solid borosilicate glass. Level-to-level bus lines need to extend through these spacer layers. Each spacer layer is a simplified version of a graphene ring plane. There are several advantages to using spacer layers: 1) they will act as heat sinks between the ring planes; 2) they will admit ambient light for the LEDs; and 3) they will make excellent invisibility planes.

Cloaking technology is already a reality, but one day it will become practical. It doesn't make sense to cloak column segments in a highly populated neighborhood, because the light bending will only reveal the clutter. But it makes total sense to make the spacer planes, which are already empty, into invisibility planes. Since there's no logic in the spacer planes, their entire function can be reserved for invisibility: each spacer plane would be occupied by 512 cloaks (see final design in Chapter 18) – one cloak for each bus column. A height ratio of two-to-one would allow the bus-columns to be cloaked in the two-high planes, while the ring planes would occupy the one-planes with no cloaking. This setup is shown below.

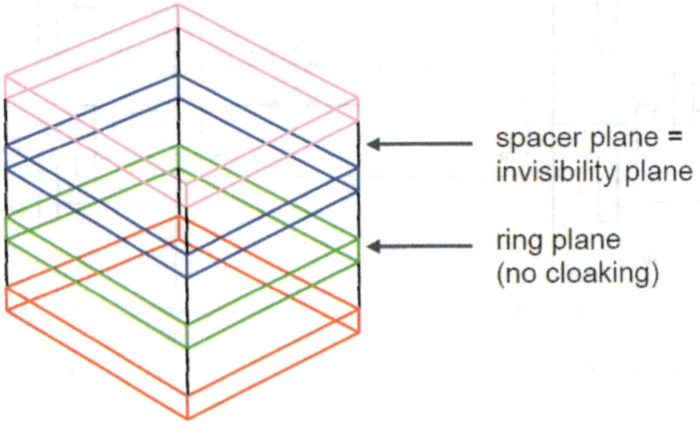

Cloaking would make the spacer planes transparent to ambient light. For voxels that are "off", the graphene in the ring planes would also be transparent to light, so that a 3D image of hollow cylinder, viewed through the center from one end to the other, would reveal the room on the other side exactly as if looking through clear glass.

The 3D display cube would employ 384 planes times 512 cloaks, or a grand total of 196,608 cloaks! For this to be practical each cloak would have to be as inexpensive as an active matrix LED element. It's not such a big number when you think of a 2D 1024 x 1024 display, which has 1,296,000 elements.

12.7. Glass Bonding

Glass melts at a temperature of around 1500 C. Large telescope mirrors, such as the one made in 1992 at the Steward Observatory in Arizona, start with clear blocks of optical glass carefully placed in a mold.

The mold is inside of a furnace, and after being melted in place, the glass is annealed for several days at a lower temperature, around 1000 C, to prevent divitrification and tension cracks as the glass cools.

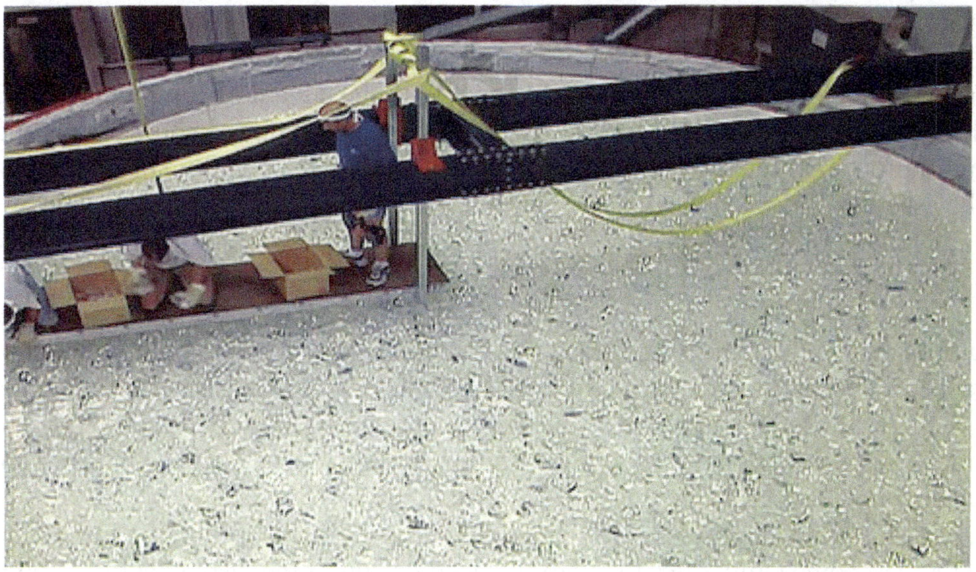

Unlike a 6.5 meter telescope mirror, the display cube is only a few inches across. Just as telescope mirrors today are ground and polished to a fine finish under computer control, a computer controller can also be used to fuse the fine layers of the cube. Subjecting the entire cube to temperatures of 1000 to 1500 C would melt the circuitry and turn the cube to slag. Using pure quartz for the layers would be even worse, since quartz melts at a higher temperature, around 1700 C. Rather than subjecting the entire cube to intense heat, a computer-controlled laser can be used to selectively fuse successive layers of the cube.

Today laser-etched glass blocks are created by using a laser that can change its critical focus under computer control – the focus in effect giving one coordinate of the Cartesian system, say "Z" – while being moved under computer control in the remaining two directions {X, Y}. The focus is strong enough and precise enough to etch an interior point of a glass block without affecting the glass around it.

In the same manner, a laser can be used to fuse the levels of the 3D cube. Instead of vaporizing the glass as in the etching process, the laser would operate at a temperature that would fuse two layers together along a very fine boundary. The laser would be turned off as it passes a row of bus columns, so that, after making two complete passes in two orthogonal directions, each ACF contact cluster would be neatly surrounded by invisibly fused glass. This kind of precision can only be achieved under computer control. Since the bus columns are otherwise embedded in the wafers, the unfused areas around the ACF contact clusters would be limited to the thin plane where the layers meet.

12.8. Conclusion

This chapter has looked at level design. The transmission line doesn't take the form of a rod. Chip-on-glass technology and anisotropic conductive film (ACF) permits layers to be stacked. Still, there are other details involved that will be covered in the next chapter. In general the LED array-elements will dominate the footprint of the ring planes.

This chapter has also looked in detail at the interface between the base plane, the bus stations, and the columns. Each bus serves four rings instead of one. At each quad nexus, the bursts are arranged into two *column-groups*, A and B, and this pattern repeats itself for all of the bus stations. Meanwhile, the optical planes remain independently clocked, though running in parallel on columns. Recall that each optical plane has, at the optical interconnect, a clock recovery circuit.

As well, the problem of fusing layers seems tenable now with the advent of computer-controlled laser technology. These lasers can be used to fuse the wafer stack into a seamless whole by applying heat selectively. This approach protects the cube circuitry and would be the penultimate step before the cube comes off of the assembly line.

References

[1] NXP, *Chip-on-Glass (COG) for LCD Modules*, www.nxp.com

[2] Online datasheets describing high-speed 8-bit ECL registers with serial-in/serial-out data ports. Clever control tactics include: shift-clock override of the input; debouncing master/slave latches; and a shift-clock suspend override. These kinds of tactics would be needed for high-speed latching of serial data in the cube's base plane.

[3] *Low-voltage differential signaling (LVDS)*, en.wikipedia.org. A description of a low-voltage high-speed serial standard that has some similarities to the high-speed design of the display cube. TI application note

SLLA108A discusses multipoint LVDS, including line-lengths, signal reflections, and impedance matching.
[4] TI, *AN-1807 FPD-Link II Display SerDes Overview*, SNLA102B. This is an application of LVDS used for sending video data from motherboard memory to an LCD over a single twisted-pair transmission line. Hints are given for {R,G,B} encoding and decoding for 24-bit applications, data rates, serializing and deserializing, and data recovery techniques. Unlike the display cube, FPD-Link data transmission doesn't use optical coupling. The transmission is through copper wires, e.g. a low-cost cable that fits through a notebook hinge assembly.
[5] Florence, Ronald. *The Perfect Machine: Building the Palomar Telescope*, Harper Perennial, 1994

Chapter 13

3D Computing Ideas

13.0. Introduction

In this chapter we continue to look at how the cube is layered together. The cube consists of horizontal planes which alternate between color layers and spacer layers. Signal lines travel vertically from the base plane up through the layers, making contact with each horizontal plane with the help of an adhesive called Anisotropic Conductive Film (ACF).

Perhaps it's telling that the reference paper that describes ACF has gone missing from the Internet. I don't think it's because the technology is top secret: more likely it was removed because the technology didn't meet expectations. The main concern with ACF is its effectiveness as an electrical conductor.

However the 3D Computing community offers an alternative. 3D Computing improves upon Moore's Law by using stacked planes to increase the transistor count. The columns that travel through the various planes of the 3D cube are described in the 3D Computing world as TSVs, or Through Silicon Vias. The technology was developed by IBM in particular [1, 2], and also by Intel, Samsung, Tezzaron, and others. However the state of the art in TSV has yet to be achieved: 3D Computing is still in its infancy. Nonetheless, like all the other technologies explored in this book, 3D Computing provides an informative background for adding substance to the 3D display-cube design.

This chapter is an overview of 3D Computing technology and its tradeoffs. There are relatively few 3D Computing papers online, probably because manufacturing CMOS wafers is difficult enough without stacking them. The big challenge in the 3D Computing world is to pack as much into as small a space as possible. The grid layout for the 3D volumetric display is also at a space premium, but is a little more relaxed in terms of transistor count, as most of the real estate is taken up by color rings.

There's a nice parallel between the 3D Computing basic design of having the CPU on the bottom layer, adjacent to the heat sink of the chip package, and memory units stacked above it; and the base plane of the display cube with all of its gating and high-speed switching, and the voxels which are the equivalent of memory elements above *it*. Both designs have all of the computing logic on the ground floor, and memory elements (basic or visual) stacked above. Both types of memory have a lower power dissipation than the CPU or base plane, but the voxel elements will have a greater cooling requirement than basic memory because light emission runs hot.

13.1. 3D Computing Examples

Below is a computer rendering of a design from Stanford. According to the blurb, the illustration represents the four-layer prototype high-rise chip built by Stanford engineers in 2010. The bottom and top layers are logic transistors. Sandwiched between are two layers of memory. The vertical tubes are nanoscale electronic "elevators" that connect logic and memory, e.g. signal paths.

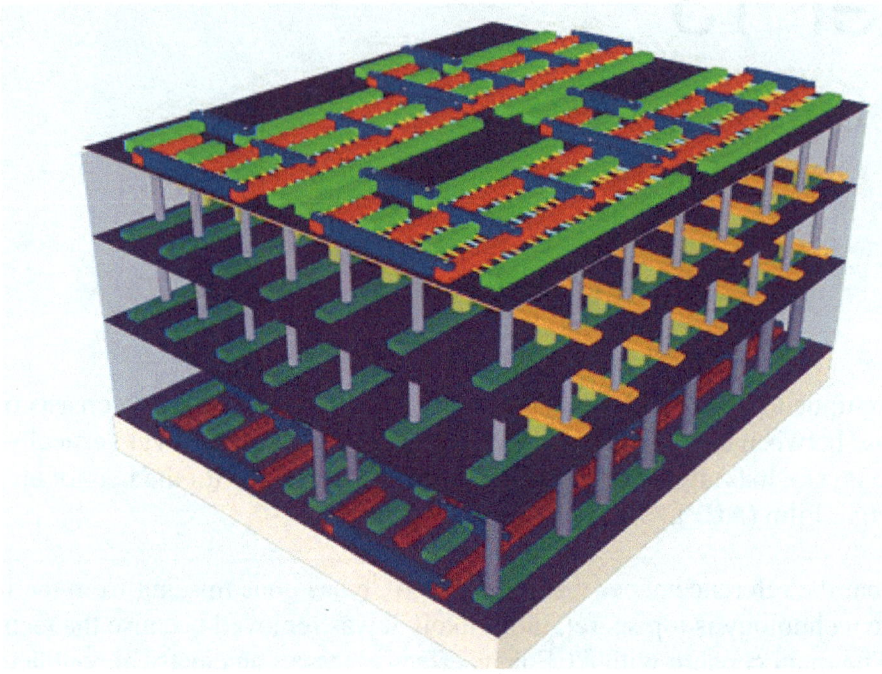

Credit: Max Shulaker, Stanford

The number of so-called elevators, or columns or TSVs, are on the order of what one might expect from the 3D display cube. There are two ways to build the layers: a *vias-first* approach creates the vias in existing or separate layers as part of the wafer manufacture, and bonds the wafers with the vias in place. A vias-first approach has the requirement that a *redistribution layer* is formed on the top of the fully processed device wafer. A redistribution layer can take the form of *landing pads* which distribute the signal paths to the new layer. A *vias-last* approach etches the vias away as a post-process, after the CMOS assembly is finished. In this case wafer bonding precedes interstrata connection creation with TSVs. Landing pads also play a role in this approach.

Below we see the steps for a vias-last approach. The vias-last process in [3] begins with a 300 mm diameter device wafer with copper landing pads integrated with the logic, the wafer temporarily bonded to a glass carrier for mechanical support. The back side of the silicon device wafer (light green) is thinned by grinding and then polished smooth by chemical mechanical planarization (CMP). The TSV is imaged in photoresist (red) and etched through the thinned silicon layer. The figure below shows: 1) the completed front-end wafer; 2) the temporary bonding of the wafer and backside thinning; 3) vias-last lithography; 4) vias-last etch; and 5) the TSV fill and de-bonding from the carrier. The aligned TSV structure must land completely on the metal pad (dark blue).

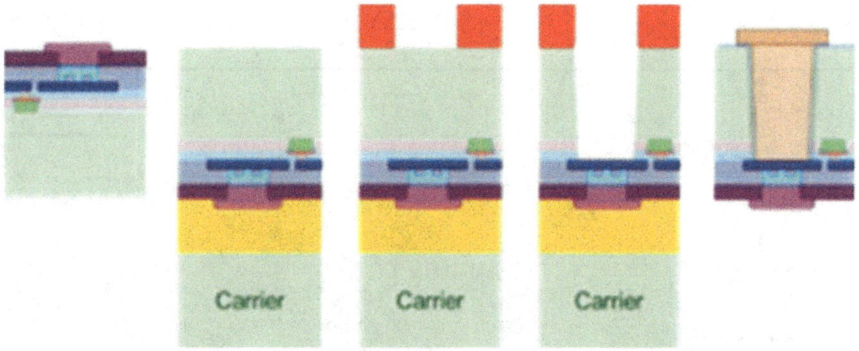

(Courtesy [3])

The conducting metal in this case is copper. Below we see something that looks a lot like the layout of the color rings on the ring planes of the 3D display device. Actually, these are four TSVs for a copper-copper contact design. The center part is etched away with chemicals and "metallized," e.g. filled with copper which makes the contact when the layers are bonded. The initial wafer contains no metal, so it can be manipulated and planarized easily. This is a vias-first plus bonding approach.

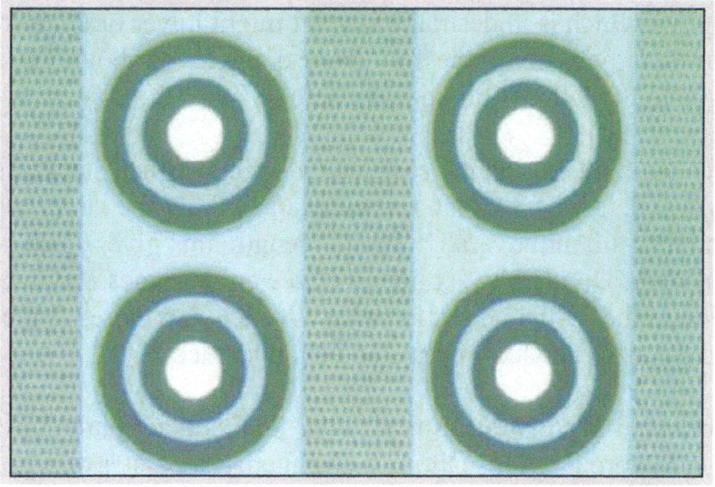

(Courtesy [1])

The center via is isolated from the rest of the silicon substrate by etched rings in the wafer around the TSV, as shown above in dark green. These rings are filled with a polysilicon (a dielectric). The dielectrics are embedded before CMOS front-end-of-line processing starts and since the rings aren't metallized they can be run through regular CMOS substrate processing. Once out of the CMOS soup, but before bonding to a device wafer, the silicon at the center of each island is replaced with metal.

A side view as seen with an electron microscope is shown below.

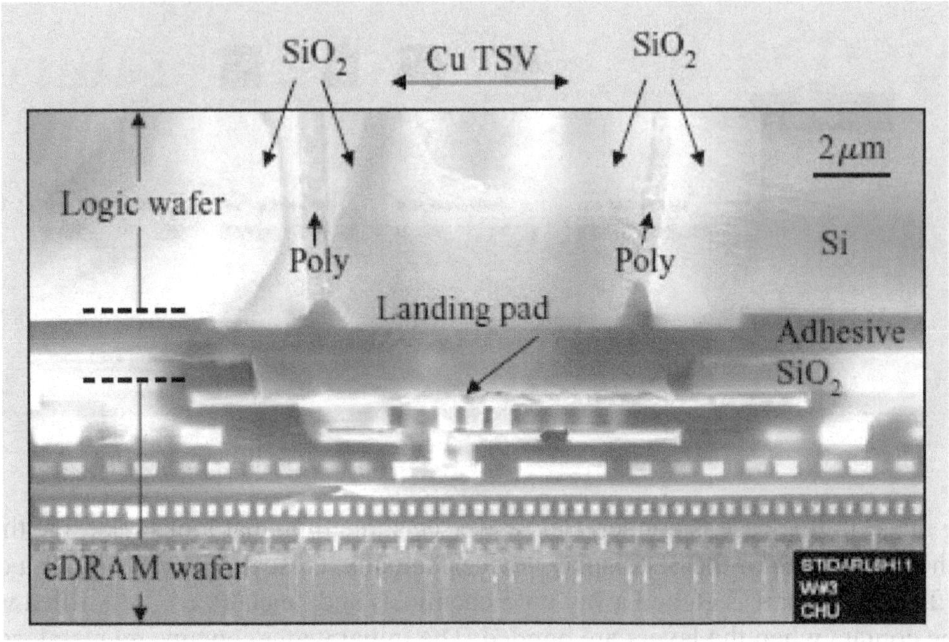

(Courtesy [1])

This photograph shows the copper "landing pad" for the TSV. Notice that the TSV itself has a void in it (top middle of photograph) – which is undesirable since it might lower or completely eliminate the via's conductivity. The scale is extremely small: the TSV has a diameter of about 5 micrometers; but notice how large the TSV is compared to the DRAM lithography below.

For the 3D display device, both the horizontal scale and also the vertical stacked scale are important. Typically a via has a vertical height of 30 to 50 micrometers. Taking the larger measurement, there are 20 layers per millimeter, and for a 60 mm overall cube side length, this gives 1200 layers that involve some kind of conducting technology. Divided by six planes (three color planes and three spacer planes), this gives an overall resolution in the vertical direction of 200 voxels. So there's a nice concordance between 3D Computing sizes and the sizes needed for the 3D display device.

Next, we see 19 TSVs arranged in a pattern similar to the ACF contact-pattern presented in Chapter 12. This seems to be a vias-last approach. The pattern is arranged as a cluster of 5 μm vias, further divided into four outlying subgroupings and one central one, rotated 45 degrees to avoid confusion with adjacent TSV device patterns during alignment. [3] The thinned silicon layer thickness is 50 μm, which is also the height of the TSVs. The lefthand image is focussed on the target resist pattern, outlined with a red ring, which is at a relative depth of 0 μm, e.g. on the surface of the silicon layer. The righthand image is focussed on the underlying device layer at a relative depth of – 50 μm. The authors of [3] have made a circular landing pad, for test purposes, for all 19 TSVs. These pictures were taken with an alignment camera with a very short focal depth, so depending on what's being imaged, the surface or the underlayer will always be out of focus.

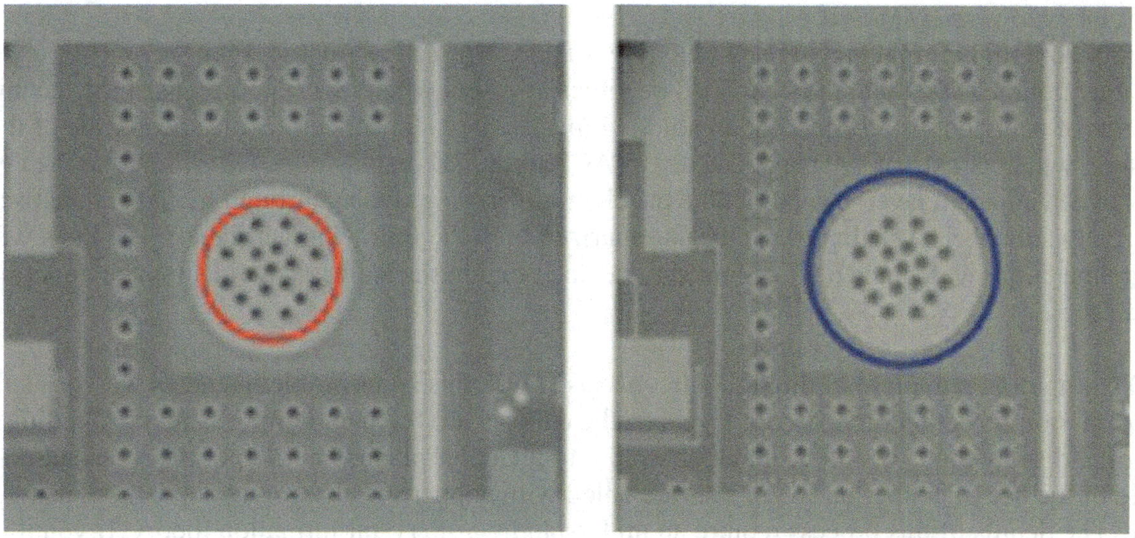

(Courtesy [3])

Obviously in a non-testbed environment, each TSV would have its own circular landing pad. However, the number of TSVs and their proximity to each other in a given layout is constrained by the error of accuracy in the alignment of the layers. The diagram below shows the relationships between various layout parameters for adjacent TSVs.

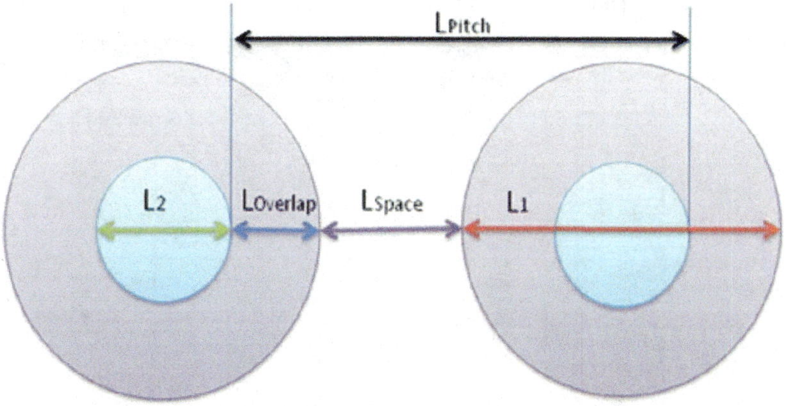

(Courtesy [2])

The light blue is the TSV and the purple is the landing pad. The *overlap* is based on the mean wafer-to-wafer misalignment, which has been determined in advance. Obviously the accuracy of the alignment process determines this number. The space between landing pads will be mostly dictated by the device design itself; but will also be constrained by the thickness of the dielectric around the via. A metal-filled TSV needs a dielectric layer for sufficient electrical isolation from the surrounding silicon substrate. The dielectric can be etched as separate rings (see image above), or it can be deposited as a coating on the etched TSV cavity itself. Process requirements for the dielectric layer, whether etched rings or a coating, include: uniform coverage; no leakage current; low molecular stress; a higher voltage-breakdown coefficient than the surrounding substrate; and immunity to bonding and annealing temperatures. Dielectric materials for coatings include SiO2 (silicon dioxide) and Si3N4.

13.2. Copper Via versus Gold ACF

As mentioned in the introduction, the conductivity of Anisotropic Conductive Film or ACF, though fine in theory, may not measure up in practice. This becomes apparent when investigating all of the technical difficulties of using copper as the TSV conducting metal. The main concern is the sheer scale of the processing, which is extremely small. The use of ACF in the 3D display cube was limited to the contact plane between adjacent layers: so in fact the problem of carrying signals through the layers themselves wasn't really solved. The challenge can be summarized with the following verbiage: "The lines will need to penetrate the entire wafer, and also will need to be anchored to the wafer to avoid slippage under heat pressure."

The 3D Computing paradigm of through-silicon-vias solves this latter problem, and as well is a solution for the transmission lines that run between layers of the cube. The TSV paradigm uses metal "landing pads" that, for the display cube, would be manufactured into the color plane itself. Paper [4] is a state-of-the-art example of the use of copper in TSVs. Vias are filled with copper via electrolyte deposition, using a $CuSO_4$ mixture. The details of this process require an understanding of crystal formation theory. If you think it's easy to fill a 5 μm via with copper, think again. There's a complicated set of parameters that governs copper deposition, the main factor being pulse-current frequency. The image below shows the effects of four different pulse-current frequencies, applied over a 30-minute processing time. The frequencies from left to right are: 889 Hz, 1481 Hz, 2222 Hz, and 4444 Hz. The highest pulse frequency has the best fill results.

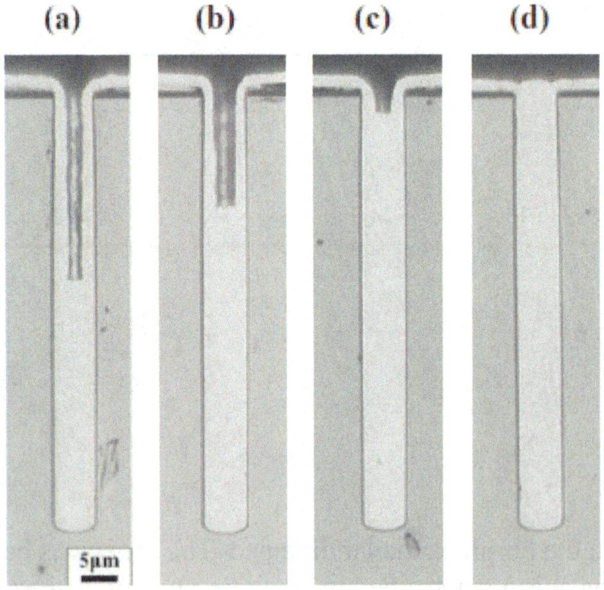

(Courtesy [4])

Electro-deposition proceeds from the bottom up, and at the same time copper is deposited on the side walls. The bottom-up filling rate increases with the pulse frequency, but the side-wall growth rate diminishes.

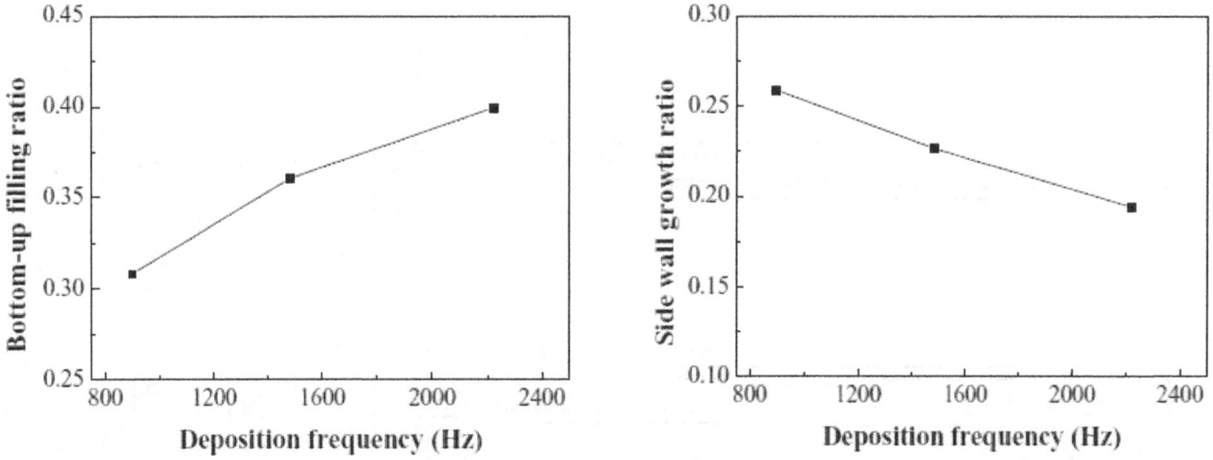

(Courtesy [4])

Besides the fill rate, an equally important goal is to avoid gaps in the fill, which can drastically reduce the conductivity of the via. The average current density (ACD) of the electro-deposition can be adjusted in this regard, and the best ACD is $2ma/cm^2$, applied over 30 minutes. [4]

It's important to realize that electro-deposition only gets the micro-crystals into the vias. Once filled, the metal must be annealed. Thus, not only filling the vias, but making sure the material is suitable for annealing, both play into the various parameters of the electro-deposition process. In the latter regard, a higher pulse frequency induces strain patterns in the CuSO4 material, and this encourages a uniform annealing without the irregular (vertical) extrusion of grains. Also related to uniform annealing is the creation of a higher density of nanoscale crystal twins when a pulse current is applied (as opposed to a direct current without pulsing). The authors of [4] annealed in a tube furnace, for 20 minutes, at 430 C, in a mostly argon gas atmosphere. The reader is directed to [4] for details and fascinating SEM photos.

13.3. Alignment

Before bonding of the wafer stack, one final thorny issue must be addressed: alignment and registration. This has already been hinted at when describing vias and their landing-pad tolerance in the form of the overlap parameter. There are many different techniques for alignment, and more arriving every day. See [2] for an overview of most of them. For the purposes of the 3D display device, as multiple wafers are stacked upon each other, it will always be a "back to front" operation. In other words, the front of the base plane will be aligned, from the back, to the marks on the front of the first spacer plane; and the front of the spacer plane will be aligned, from the back, to the marks on the front of the first color plane; and so on, with the the spacer and color planes alternating all the way up to the cube top. But how is this done?

The most appealing alignment technique uses stepper-motor-driven stages under digital control. It's called the SmartView Alignment Method™. A pair of microscopes (one on the "top" side and the other on the "bottom" side of the wafers to be aligned) are placed so that they don't interfere with the digitally-controlled placement of the top and bottom wafers (which are on stages). The wafers to be aligned are placed "face-to-face" with a gap of less than 100 μm and vacuum-sucked to the top and bottom wafer stages, respectively.

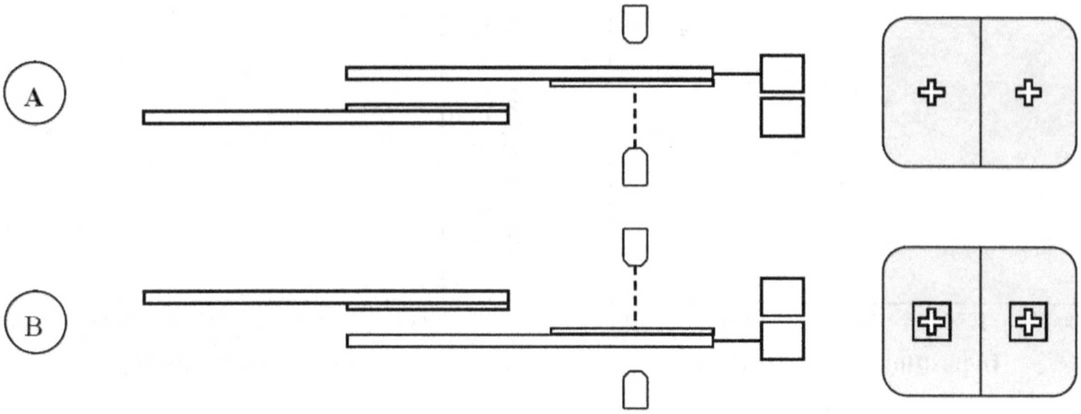

The two wafer stages can be moved back and forth horizontally. After the top and bottom microscopes are aligned with each other (i.e., calibrated), the top wafer stage is moved in between the microscopes; the objective-mark in the bottom microscope is aligned with the alignment mark on the top wafer. (Step A above.) The top wafer position is digitally stored, and the top stage is retreated.

In step B, the bottom wafer stage is moved into position against the top microscope. The alignment mark on the bottom wafer is then aligned with the top microscope. *The top wafer stage is then moved back to its digitally stored alignment position.* Finally, the two aligned wafers are vertically moved to contact each other and are clamped to hold the alignment for wafer bonding. Misalignment tolerances can be achieved on the order of 0.25 μm for this method. However the "face to face" orientation makes this method unsuitable for a wafer *stack*.

The SmartView system is an extrapolation of the "same side" alignment paradigm which only requires a single microscope on the top. The "same side" can be either the front face or the back face, but it must be one or the other for both wafers. Assuming a frontside approach, the front of the first wafer is aligned with the microscope, and the position is digitally stored. The second wafer is moved between the first wafer and the microscope. The alignment mark on the frontside of the second wafer is aligned with the *digitally stored image* of the alignment mark on the front of the first wafer. The wafers are vertically joined and the process can be repeated for the next wafer, using the top of the stack as the new reference, although care must be taken to avoid *accumulated drift*. The misalignment tolerance in this case is ±5 μm, but can surely be improved.

Basic alignment depends on the accuracy of the lithography of the alignment marks: if the marks are off from wafer to wafer, then all bets are off. There's another technique for alignment based on etching grooves or footings into the substrate. This isn't as desirable as the digital technique for two reasons, both pertaining to the display-cube application. First, it will be hard enough to align flat-surfaced wafers and at the same time achieve invisible joins, since the display cube must look monolithic even though it's composed of hundreds of wafer layers. From microphotographs of the tongue and groove approach, neither the cutting nor the final alignment give one confidence. The display cube can have no gaps between individual layers due to sloppy mechanical grooves; such gaps will create microsurfaces that reflect ambient light. Second, mechanical grooves leave no room for play. Once two layers are "seated," if the mechanical alignment is off, it's off for good. On the other hand, a flat surface that uses two-dimensional marks leaves room for play, so that an overall acceptable alignment can be achieved even if some of the individual alignments are off, though still within the acceptable overlap tolerance for the landing pads. Below are two mechanical alignment designs, also called *keying structures*, but neither are acceptable for the display cube for the reasons just given.

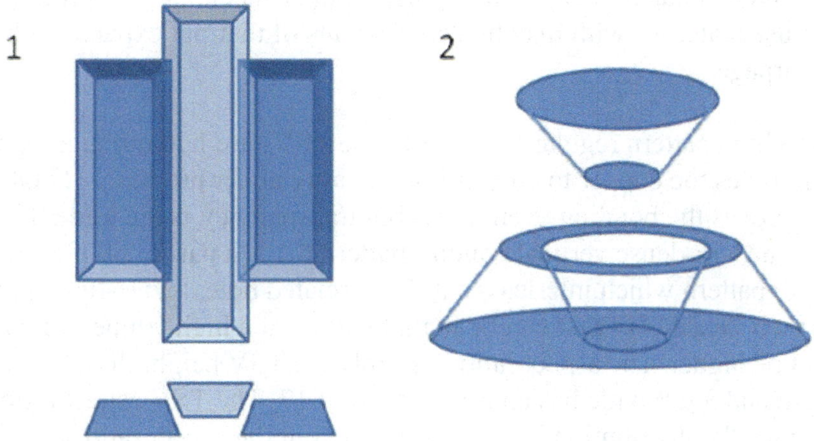

(Courtesy [2])

13.4. Bonding

Besides the normal overlay error of the standard CMOS lithography steps, which might affect the accuracy of alignment marks on the wafer, bonding of the wafers may introduce additional alignment error.

Bonding comes after the device wafers have been manufactured, and for the vias-first approach, after the TSVs have been etched, filled, and annealed. State of the art for bonding systems gives an alignment error margin of about 1 μm. Bonding error must be figured into the landing-pad overlap margin along with all of the other factors. One big factor is the warpage of the wafer itself during bonding. Another factor is temperature. CMOS circuits are made with back-end-of-line (BEOL) metallization, which exists prior to the bonding, and presumably the TSV metal will have already been deposited and annealed as well. The existence of CMOS metal means that the bonding temperature must be less than 400 C. This is cooler than the copper annealing temperature used by Samsung of 430 C.

One can imagine that the initial spacer planes contain no metal prior to the filling of transmission-line conduits with copper; and can be planarized to very fine tolerances. TSVs will come into play in the *distribution layer* that feeds signals to each color plane. After being etched, each distribution layer can be filled and annealed; then planarized again to remove what Samsung calls "overburden layers". These distribution layers are then bonded with the various color planes, the latter outfitted with landing pads on the bottom of the wafer. The surface topography of each color plane is already so complex that adding additional surface topography for the TSV landing pads shouldn't increase the complexity.

Copper bonding, which is what we're interested in, works by applying pressure to the wafer stack at elevated temperatures. Even though the copper has already been annealed in the TSV itself, the actual contact between the TSV and the neighboring wafer takes place during bonding. The bonding quality can be controlled with: a) surface preparation techniques; b) thermal annealing cycles; and c) the use of optimized pattern geometry.

In [2] the authors found that wafer bowing (warpage) increases to 80 μm as the bonding temperature rises to 460 C, but that bowing is only 10 μm at 330 C. Alignment accuracy is within ±5 μm when the bow of silicon/glass anodic bonding is 10 μm. This isn't great, since the TSVs themselves are typically 5 μm in diameter. However, bow control can be achieved by using 7070 glass in combination with silicon. 7070 glass bows when the bonding temperature reaches the strain point, but in the opposite direction to the bow due to thermal expansion mismatch. Ideally 7070 glass would cancel out the induced warpage. However the best practice is to use materials with matched coefficients of thermal expansion, because mismatched coefficients induce warpage.

The display cube has a high pattern regularity – that is, the TSVs are laid out in a regular cellular array – and this regularity improves the copper to copper bond quality under pressure. [1] Obviously, increasing the pattern density improves the bonding even more, but transparency is the tradeoff here. Still, there's a nice synergy between the less-dense vertical conduit pattern that's a part of all planes; and the denser color-plane/distribution layer pattern which interleaves it. On a related note, a vias-first approach accommodates a high density of TSVs because vias-first can be manufactured at a high temperature and thus the aspect ratio of each TSV can be higher. The aspect ratio is simply the TSV height divided by its diameter. Thus a TSV that's 50 μm high and 3 μm wide has an aspect ratio of 17. The TSV aspect ratio in each distribution layer would be low, since the distribution layer is very thin; but the aspect ratio would be higher for the spacer plane and color plane interconnects.

One other concern is whether to bond the entire cube all at once, or layer by layer. Here the constraint is temperature. Two objections can be raised. First, the all-at-once approach would require the entire stack to be prealigned, thus putting a bigger burden on the pre-alignment process. Second, heating the entire cube makes an assumption that the bulk of the cube has no effect on bonding temperatures being achieved in the interior of the cube. This is a false assumption. Therefore a layer-by-layer bonding approach is the only option.

More precisely, TSV bonding between a color plane and its distribution layer is treated as a separate step in the manufacturing process. Once these two components are bonded, they're treated as a unit when aligning the rest of the stack. Recall that the spacer planes are interleaved with the color planes. Each layer is allowed to partly cool before the next layer is fused into place. Spacer planes can also act as heat sinks when they're in the next-to-top position and the color-plane assembly is fused on top.

According to [2], intimate contact between two wafers is essential to equalize the temperature of the wafers. A 1 kN (kilo-Newton) downforce is required to bring adjacent wafers into intimate contact and to preserve alignment. As well, a slow ramping rate (6 C/min) avoids the large mismatch of the thermal expansions between wafers by keeping the temperature difference between wafers within 2 C. The profile that works the best seems to be a small prebonding force of 1 kN, a slow ramping of 6 C/min, and a full bonding force of 10 kN at 400 C for 1 hour.

Finally, the use of adhesives for bonding is discouraged. Copper-copper bonds between the layers have excellent thermal conductivity because metallic connections between the wafers provide the best means for heat transfer. Intermediate adhesive layers will decrease the thermal conductivity between layers. As well, adhesives will cloud the transparency of the cube. So we now say goodbye to ACF. Instead, TSVs will be used for vertical transmission lines. The junctions which make contact with every single layer will be bonded as described above, no glue required.

13.5 Stiction

The next step after the entire cube is bonded is to fuse the layer boundaries using the laser technique described in Chapter 12. Before lasing begins the cube is held together by *stiction*. In other words, the layers are aligned and they're bonded and the metallic joins distributed throughout the cube hold it together by stiction.

However the layers are not completely fused: air gaps may exist at layer boundaries no matter how good the planarization. The computer-controlled laser will eliminate any air gaps as it selectively fuses each boundary between planes. The cube is held tight during the lasing operation, but stiction already holds it together to some extent. Once the lasing is finished, the display should appear as a solid, optically clear cube of glass.

13.6. Conclusion

This chapter has taken a close look at alignment, registration, through-silicon-vias (also called *through-strata-vias*), and bonding and annealing. The information has been drawn from the 3D Computing community, which includes players like IBM and Samsung. There are quite a few similarities between the mainstay 3D Computing application of a processor with memory planes stacked above it, and the cube's base plane with color planes stacked above it. As well, the dimensions of the two applications in the circuit plane are about the same.

The main difference is that current 3D Computing designs are limited to about four levels. The display cube has 384 preassembled color planes and 384 spacer planes. Another difference is that the display cube is visually transparent, or should be; while the visual constraints of a regular 3D Computer aren't so important since the stack is hidden in a "package".

This chapter has also looked at a few digital alignment techniques which, with a little imagination, can be extended to an automated alignment tool for a stack of wafers, rather than just two. "Same side" and also "face to face" alignment can be achieved with stages driven by very fine stepper motors with offsets that are digitally controlled. The "eyes" of the system are microscopes that produce images that are also digitally stored. This technology already exists.

As well, the technology of through-silicon-vias, or TSVs, has been presented. This seems to be the most exciting aspect of all of this "back room" technology as it pertains to the display cube. A vias-first approach is the only feasible option for the display cube, and copper-on-copper bonding is also a good match for the cube because the bonding improves when the TSV patterns are regular or equispaced.

References

[1] Koester et al, *Wafer-level 3D integration technology*, IBM J. Res. & Dev. Vol. 52 No. 6 Nov. 2008, online

[2] Sang Hwui Lee et al, *Wafer-to-Wafer Alignment for Three-Dimensional Integration: A Review*, Journal of Microelectromechanical Systems, Sept. 2011, online

[3] Flack and Slabbekoorn, *Overlay performance of through silicon via last lithography for 3D packaging*,

2018, online
[4] Jin et al, *Through-Silicon-Via (TSV) Filling by Electrodeposition of Cu with Pulse Current at Ultra-Short Duty Cycle*, Journal of The Electrochemical Society, 160 (12) D3300-D3305 (2013), online

Chapter 14
Quantum Dots

14.0. Introduction

Samsung's Advanced Institute of Technology (SAIT) in South Korea has been working on next-generation display technology for its product lines. The latest generation of consumer big screen televisions uses quantum dots, and Samsung's consumer literature gives this definition:

> *A Quantum Dot can be defined as a crystal, like a diamond, that turns light into color.*

Pure marketing of course. As we will discover in this chapter, the reality is far more complex.

In nanotechnology terms, a quantum dot is indeed a crystal growth in a hexagonal form, otherwise known as a photonic crystal, and its property is to be a *single emitter*. [1] A lattice of single emitters will produce a far-field effect, and depending on the size of the quantum dots, light will be emitted at different visible frequencies.

For LED displays, quantum dots are used as *luminophores*. These luminophores are driven by electricity gated through thin film transistors (TFTs). Just as ink is printed onto paper, quantum dots can be *transfer-printed* onto a *display plane*, and then the display plane can be activated by TFTs that reside on a backplane, behind the display plane. [2] This is shown schematically below.

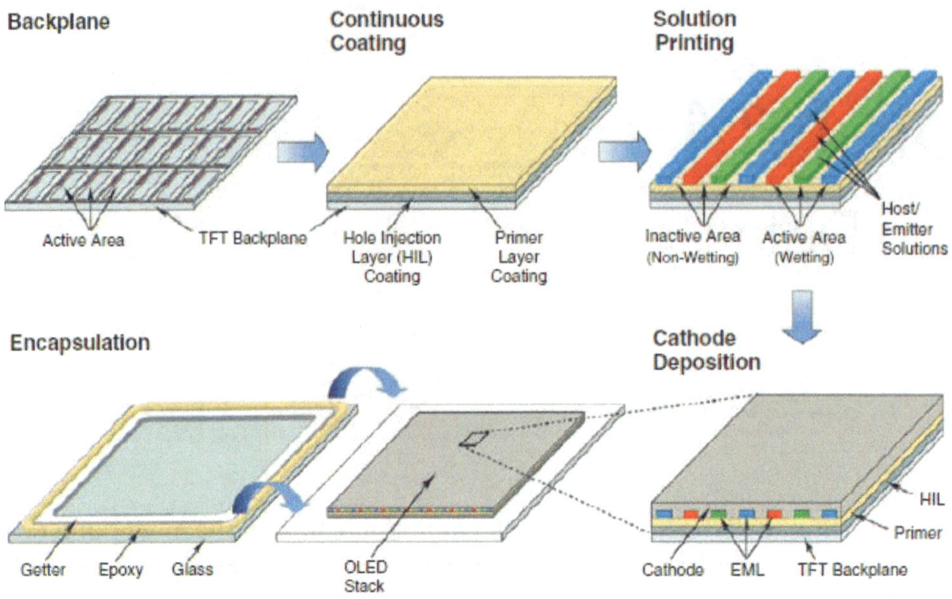

An LED screen with a cathode-top design. (Courtesy [5])

As will be shown, the overall quantum LED product is a marriage between two very important technologies: quantum dots and TFTs.

14.1. Colloidal Quantum Dots

Professor Jong Min Kim of the Department of Engineering, Cambridge University, is one researcher working on quantum-dot-based technology for next generation lighting and displays. *Colloidal quantum dots* (cQDs) are processed using self-assembly techniques, so that the array of dots forms a regular and close-packed two-dimensional matrix. In fact, quantum dot arrays as used in QLEDs have depth as well, usually about five layers, and the depth direction also shows some level-to-level regularity.

Professor Kim's team demonstrated the first backlit display devices in 2006 and later in 2010 using cQDs in photo-luminescent (PL) mode; and then demonstrated in 2011 and 2013 cQDs in electro-luminescent mode (EL) in an active-matrix configuration. A cQD backlight unit for LCD displays went to market in 2015 under the brand name of Samsung SUHD TV. Quantum-dot displays went to market in 2017 under the brand name of Samsung QLED TV, with improvements released in 2018 and 2019. However, customer reviews revealed that the quality assurance on these new QLED TVs wasn't 100 percent. Defects were found in the early models, such as darkened regions on the TV borders; or in one case a rectangle that became dim because the electrical drivers somehow malfunctioned (probably a subregion driven by its own TFT array). We'll see that this isn't surprising given the complexity of the QLED technology.

Photo-luminescent (PL) cQDs offer superior brightness compared to traditional phosphors and organic LEDs (OLEDs). Additionally, cQD-based lighting benefits significantly from the intrinsic reliability of the quantum dot inorganic material structure, which reduces the need for air barrier layers with low water vapor transmission ratios (WVTRs).

The investigators expect that cQD LEDs will replace current technologies through: (1) superior reliability of the inorganic structure in an almost air barrier free architecture with reference to OLED (which requires a WVTR of 10 to 6 g/m^2/day); (2) lower power consumption (60% less) and lower product cost (50%

less) than current OLED technology; and (3) colour purity of 110% or greater compared to typically 80% for OLED. Sony, a competitor of Samsung, is the main producer of OLED TVs.

Quantum dots are made of cadmium selenide, cadmium sulfide, or indium arsenide. All of these contain a heavy metal (cadmium) or arsenic. Alternatives that have been tested are indium phosphide, zinc selenide, and zinc sulfide. Unfortunately all of these alternatives suffer in the blue spectrum. Blue emitting QDs have low performance due to the smaller particle size, with more plasmon surface states compared to red and green QDs. EL performance of InP QLEDs versus cadmium-based devices is generally inferior, having a lower maximum luminance. One research direction is to grow short organic ligands around the inorganic quantum-dot core, thereby encourging uniform spacing when the QD film is self-assembled.

The QLED display plane is made by *transfer printing*. In this process a donor substrate is spin-coated with quantum dots (QDs). The uniform QD film mounted on the donor substrate is transfered to a receiver substrate. To facilitate a smooth transfer, the donor substrate is coated with octadecyltrichlorosilane (ODTS), which is like PAM used in baking to allow cupcakes to fall out of the cupcake pan easily. During spin-coating, the QD suspension is deposited onto the ODTS film and the ODTS smooths out the suspension to produce a more regular QD film. [2]

The receiver substrate is in fact the display substrate: in the display cube world this would be a color plane. But instead of printing rings, as would be done for the display cube, 2D QLED displays use printed strips of alternating red, green, and blue QD films. The transfer is a solvent-free inking and stamping process at the nanoscale level. The process is shown below.

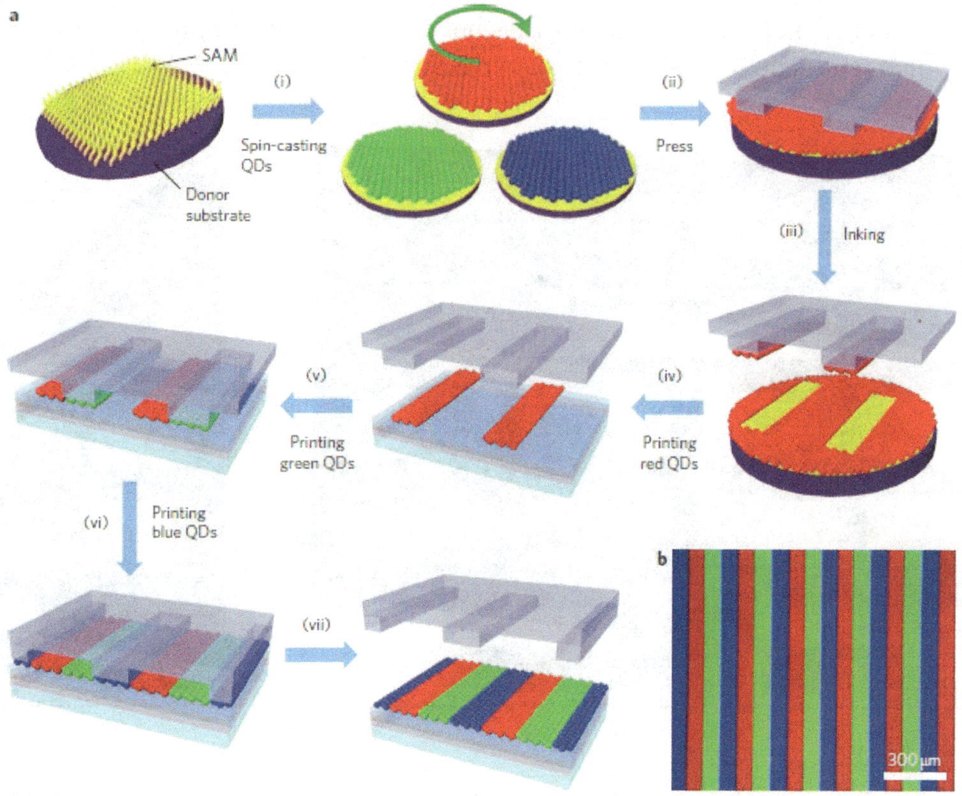

(Courtesy [2])

Monolayer deposition methods and transfer printing are associated with lower production costs. cQD researchers are aiming to achieve cost reduction in manufacturing by using continuous, as opposed to batch, cQD synthesis; monolayer resin-free processing; and all inorganic interface materials for the TFT ETL (electron transport layer) and HTL (hole transport layer, see next section).

In general, QD active displays have a higher pixel definition with higher purity and narrower emission/absorption than non-QD displays. In 2011 Samsung produced a prototype 4-inch active-matrix QD display, shown below, the reproduction not doing it justice.

(Courtesy [2])

Also in 2011 the researchers in [5] used organic LEDs but a very similar printing process to produce sixteen displays which represented a full screen, shown below.

(Courtesy [5])

Manufacturing and processing must ensure that monolayer cQD films are close-packed and almost void free. Printed optoelectronic films for EL cQD active devices consist of a cQD monolayer driven by a TFT driver backplane that itself contains several layers.

14.2. Thin Film Transistors

The TFT that seems to work best with QDs is the hafnium-indium-zinc oxide (HIZO) TFT. This is a lab creation and the hafnium makes it very expensive.

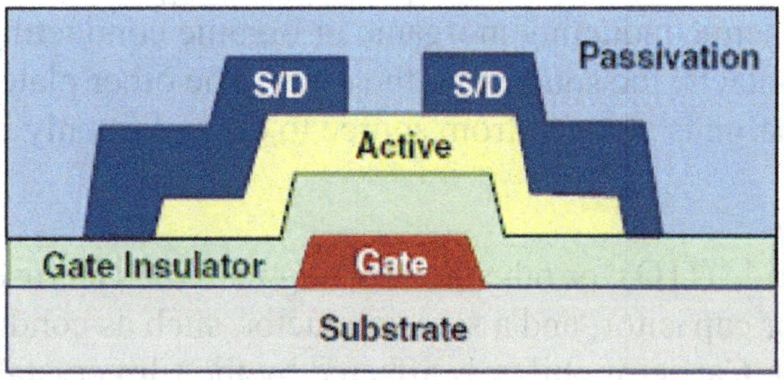

(Courtesy [3])

The stacked structure shown above can be explained as follows [3]. This is a bottom-gate, top-contact structured HIZO. The substrate (lowest level) is Corning glass. The gate electrode (red) is 200 nanometer-thick molybdenum, deposited by sputtering. The gate insulator (neon green) is a combination of silicon-nitride (presumably the inert lower layer, 400 nm) and silicon-oxide (the less inert top layer, 50 nm). These layers are deposited (e.g. "grown") by plasma-enhanced chemical vapor deposition (PECVD). The active layer (yellow) is HIZO, deposited by sputtering at room temperature. The researchers in [3] used different mixtures of hafnium in HIZO to test the electrical characteristics of the compositions. HIZO acts as a semiconductor.

Next, the contacts labeled S/D are source-drain electrodes. In [3], the source-drain electrodes are made of molybdenum, as is the gate electrode. Researchers are now experimenting with electrodes made of graphene. Graphene can also be spin-coated onto a substrate and printed much like the transfer printing process used for QD films. So there's a nice symmetry for the two processes, one for the QD films, and one for electrodes made of a graphene film.

Finally, a silicon-oxide film (200 nm thick) is deposited for the passivation layer. The entire thickness is about 200 nm x 5 = 1000 nm or one micrometer. Existing microphotographs of these TFTs or ones like them are actually rather sloppy looking, looking more like a lumpy roast beef sandwich than a close-fit assembly. Part of this sloppiness is due to the flexibility criterion. A flexible substrate doesn't really admit a rigid block structure. However the 3D display cube doesn't require flexibility: in fact, quite the opposite; we want rigidity.

High device stability depends on appropriate gate dielectric materials (e.g. in the gate insulator layer), and an impermeable passivation layer (quartz) to protect from environmental humidity. Hydrogen can also develop from the manufacturing process. Careful deposition of the gate insulator via the PECVD process, by controlling the hydrogen in the PECVD precursors, can reduce water contamination into the HIZO, which can become conductive through the incorporation of hydrogen. The HIZO must remain a semiconductor.

As well, the etch technique plays a role. The two types of etching are: a) back-channel etch (BCE), which for fabrication requires fewer photolithographic steps; and b) etch stop (ES), which requires an additional photolithographic step, but is more efficient at protecting the gate insulator layer (the dielectric) from the source/drain electrode patterning process, and thus improves TFT stability.

14.3. Theory

There are two types of transistors: bipolar and unipolar or field-effect transistors (FETs). A FET has three terminals: source, drain, and gate. A FET usually operates as a capacitor and its internal body is composed of two plates. One plate is a semiconducting inorganic or organic conducting *channel* (the active layer) with two opposing ohmic contacts: the source and the drain. The other plate (the gate) controls the charge induced into the channel. Charge is induced from source to drain, but only if the gate voltage is higher than a threshold value.

Doped silicon, indium tin oxide (ITO), or other conducting materials are used as the gate; a dielectric serves as the gate insulator or capacitor; and a semiconductor, such as conducting polymers or metal oxides, serves as the channel. Carrier mobility is affected by the film crystallinity and the nanostructure. Electron mobility is measured in square centimeters per volt per second. HIZO mobility is very high (which results in a brighter LED): 7 $cm^2\ V^{-1}\ s^{-1}$. Mobility is directly related to "high field effect" or basically to the "far field effect," e.g. discernable radiation. Mobility can be impaired by the presence of imperfections in the film nanostructure.

14.4. TFT Meets QD

TFT electrical design is very much related to the LED design that it drives. A rough schematic of the combined structure is shown below.

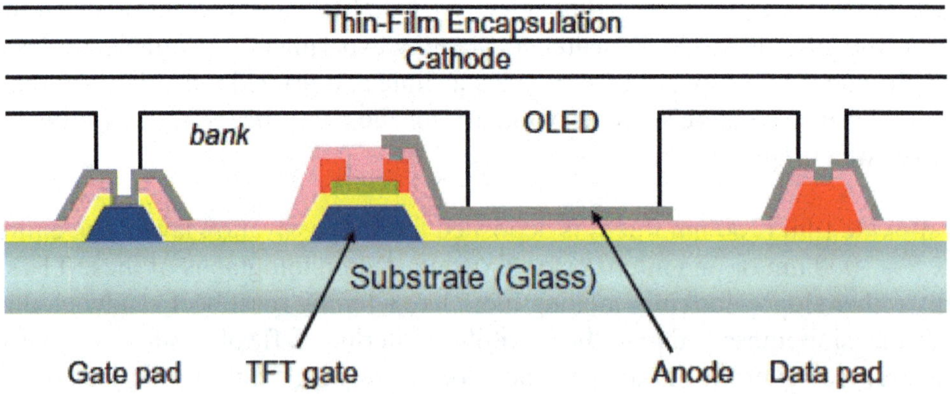

A bottom-gated TFT on the anode side of the LED stack. (Courtesy [6])

QD and OLED devices operate by the same principles, as the two are very similar. In an OLED device, thin organic films consisting of an electron transport layer (ETL), an emissive layer, and a hole transport layer (HTL) are situated between an anode, which is made up of transparent indium tin oxide (ITO, similar to HIZO) and the metallic cathode which uses a low-work-function metal such as magnesium, calcium, aluminum, or lithium-aluminum. In one paper [2], the HTL itself is made of an organic metallic polymer. Notice in the diagram above that the anode is connected to the TFT drain electrode.

During operation, the anode is positive with respect to the cathode. Recall that an anode is a positive electrode, and a cathode a negative electrode. Under an applied electric field, electrons from the cathode are injected into the LUMO (lowest unoccupied molecular orbital) of the ETL, and electrons are withdrawn from the anode. In other words, holes from the anodes are injected into the HOMO (highest occupied molecular orbital) of the HTL.

Electromagnetic forces cause charge carriers to drift and approach each other. They finally meet in the emissive layer to form the *exciton* which is the bound state of an electron and hole. Both charges may arrive on a single molecule, or two molecules may form excited states. The radiative decay of this excited state results in the emission of light. The frequency of this radiation depends on the energy difference between the HOMO and LUMO levels of the emissive material (which in turn, for QDs, depends on the QD size).

The HTL facilitates hole injection and transport from the anode into the emissive layer, and as well blocks escape of electrons (not photons) from the emissive layer. The HTL is designed to have low electron affinities and ionization potentials, to promote electron removal. In other words, oxygen in the anode's oxide material causes metal ions in the metallic polymer of the HTL to be pulled into a higher state, producing an electrophilic reaction that "abstracts electrons." Holes left in the lower levels move under the action of the electric field like positive charges, leading to "hole conduction." This makes the anode a positive-type (p-type) semiconductor.

Electrons take part in conduction by absorbing energy from the electric field, and, emitted from the cathode, they rise to a neighboring unoccupied level, starting with the lowest unoccupied orbital (e.g. the LUMO). Hence the ETL charges up negatively. This makes the cathode a negative-type (n-type) semiconductor. The stack as drawn in [6] is shown below.

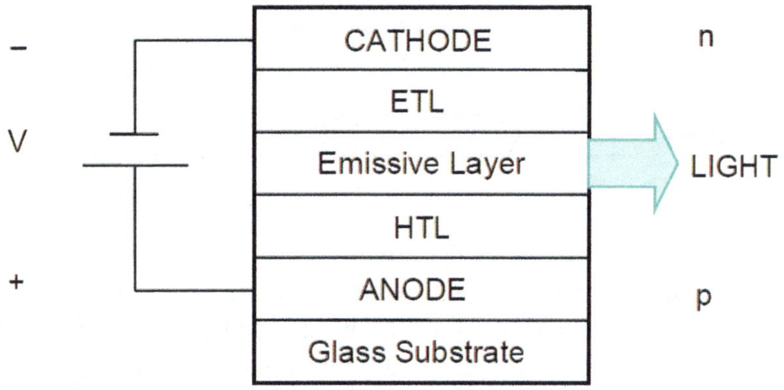

The TFT circuit that drives an individual LED is not trivial. The authors of [6] give the following circuit diagram, involving four transistors.

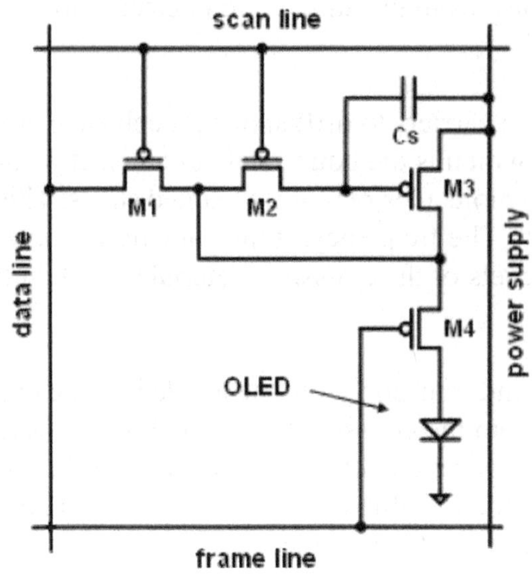

The various signal lines (data, frame, scan line) are on different layers of the OLED stack and thus don't interfere with each other. An associated signal diagram is shown below.

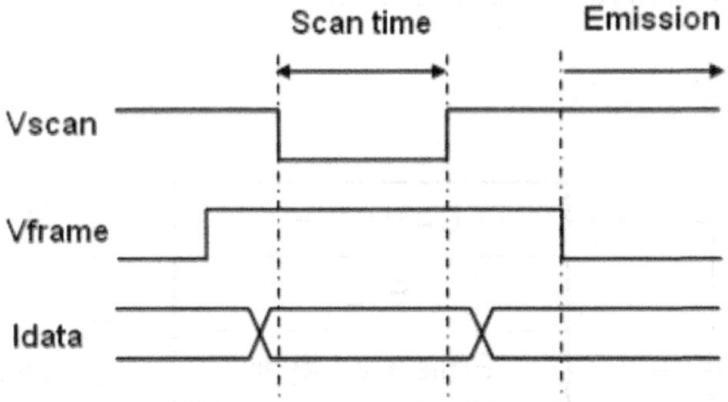

In practice, there is some variability in the threshold voltage in an active-matrix. In the diagram above, M1, M2, and M4 are *switching* TFTs. Notice that M1 and M2 are driven by a *current*, associated with the data line. This controls the illumination level of the pixel. (On the digital side, for instance, the Samsung QLED TV is quantized to 10 bits each for R, G, and B, producing over a billion colors.) The frame and scan signals are voltages, and are connected to the gates of M1, M2, and M4, the switches. M3 is the *driving* TFT, and is connected to the power line.

The signal diagram shows that the current for a pixel is gated by the scan signal, but the current isn't fed to the OLED until the frame signal goes low and gates M4.

The composition of the HIZO has an effect on voltage bias. Higher zinc ratios give a lower voltage drift or bias over the lifetime of the product [3]. But other research shows that increased hafnium ratios give the same advantage, with even better results [2]. In general the transistor works so that when a gate voltage is applied, the drain current increases with it and the charge is let through.

Thin film transistors are built up from several layers. An LED with a *bottom-gate* means that the transistors are integrated into the bottom layer relative to the viewer. In the 3D display cube world, since the cube is usually being viewed from above, the bottom-gate integration is probably best suited to the overall design. Given that the cube base is oriented as the "bottom", then the TFTs in the cube, though layered vertically, can still be thought of as, for example, "bottom-gate inverted-staggered." [4]

As already mentioned, the HTL is in contact with the QD film and is made of polymer chains. A uniform QD film surface ensures that the QD film is in close contact with the HTL. Close contact enhances the charge transport. In practice, the quantum dot film is transfer-printed directly onto the HTL layer. On the other side of the QD film is the ETL, made of titanium dioxide, which is deposited by spin-coating.

Electron mobility from the HIZO TFT backplane is such that electrons are released and "injected" into the hole layer, which then stimulates the QD layer, resulting in electro-luminescence. More or less. But researchers are looking at moving away from the polymer chain HTL, which is organic, to an inorganic material. Interface materials for emissive cQD films with inorganic HTL and ETL layers for EL devices are being designed and fabricated at the device integration step by the Cambridge research group.

14.5. Ring TFTs for the Display Cube

One of the main takeaways from the discussion so far is that an active-matrix is made up of layers. Depending on the charges and orientation, the device can be cathode-top or anode-top. For a 2D display a single layer can serve as a backplane that doesn't have to be transparent, since the display is viewed from only one direction, usually from the cathode direction. In three dimensions *everything* has to be transparent, otherwise each display plane in the vertical direction would block all the other display planes.

The color rings of the 3D display are different from pixels in that the individual TFT isn't driving a rectangular region; it's driving a circular region with a hole in it, otherwise called an *annulus*. From [6] we know that the actual printed width of TFT M3 is 575 micrometers, the largest of the four transistors, since it's the power feed for the display plane. For the ring plane, a rectangular TFT won't fit the circular layout. What's needed is a *circular TFT*.

It turns out that just such a beast exists, called a Corbino TFT. [8] The design is shown below.

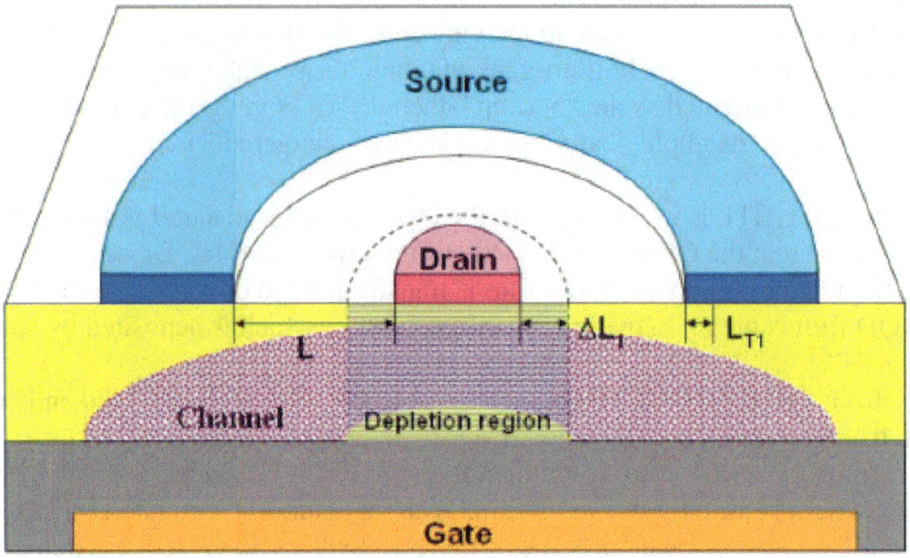

(Courtesy [8])

All of the components of a TFT are represented in this diagram: the gate, the gate insulator (gray), the channel representing current flow (pink), and the source and drain, as labeled. But instead of a rectangular layout, it has a circular layout. The Corbino can also be arranged so that the source and drain locations are swapped, but this alternative layout has some problems. The layout as shown is the optimal configuration for driving an LED display.

In general, the width to length ratio for driver TFTs is very high, for example, 100 to 1. The Corbino can attain an even higher ratio. The driving current is modeled to match observed results with a complicated geometric correction factor that uses radius and circumference. The dimension L as shown above is given as 5 to 6 micrometers, which makes the overall diameter about 22 micrometers – a very compact size but bigger than the Okidata micro-LEDs.

In short, the Corbino has a relatively small footprint, but a high W/L ratio; a high ON current; and a lower than average OFF current. The OFF current depends solely on the volume of the depletion region, shown above. Obviously, a reverse configuration would have a higher OFF current, making it unsuitable as an active-matrix driver. Note that the authors of [8] use an anode-top design rather than a cathode-top design for their LED stack.

Space is at a premium in the ring plane of the 3D device. Therefore a ring-shaped TFT should simplify the layout since the color rings themselves are the same shape. The Corbino layout introduces a new constraint, namely that the control island at the center of the color ring is now taken up by the Corbino drain electrode. In fact, a distributed layout for the contact clusters at each ring plane transmission line interface would put the control circuitry outside of the LED, so that there's no conflict.

To avoid space conflict, signal lines are put on a different layer in the printed circuit than the LED stacks. But all of these layers may have consequences as far as display-volume transparency goes. HIZO or something just as transparent may solve part of the problem. As well, there is the problem of the *sous-bande* (Chapter 11), which needs its own driving circuit.

14.6. Environmental Impact

The Cambridge group aims to explore the design and synthesis of cadmium-free cQDs with core/shell structures using continuous flow production methods which can then be incorporated into active devices. Cd-free cQD materials have been demonstrated by major suppliers of quantum dots such as Nanoco (UK), Nanosys, QD Vision (US), and Merck (Germany), who are among the major producers in bulk of Cd-free cQDs for backlight applications in LED displays. These cQDs are consumed by system integrators including LG Philips, Sony, TLC, and HiSense.

Cadmium was probably originally chosen because it naturally adopts a hexagonal close-packed structure and has been used since the 1950s in semiconductor design. Magnesium also has the hexagonal property, but because it's an alkali metal rather than a heavy metal, under high current it tends to break down.

The possibility of integration of cQD optoelectronics with current semiconducting technologies has accelerated the pace of research in the cQD area, and an increasing number of research groups and companies are currently active in this area worldwide. The hafnium in HIZO is in the same metals group as cadmium, but it's both rarer and heavier, so like cadmium, it's not a good candidate for consumer electronics.

14.7. Conclusion

This chapter has surveyed the world of quantum dots and TFTs and has outlined some of the tradeoffs in LED stack design. As well, a new kind of TFT called a Corbino* has been introduced. The Corbino can be referred to more generally as a *ring TFT*.

One important characteristic of HIZO TFTs is that they are optically transparent. This will be an important consideration when building the display cube; ITO is also transparent and is used as the metallic cathode for most active matrix stacks. The world is waiting for a semiconductor technology that: a) is transparent; b) has the electrical characteristics needed to produce the current for light emission; and c) is environmentally friendly. Stay tuned.

References

[1] Benisty, Henri; and Lourtioz, Jean-Michel. *Photonic Crystals for Integrated Communication Systems*, in *Applications of Metamaterials*, Capolino (Ed.), CRC Press, 2009
[2] Kim et al. *Full-colour quantum dot displays fabricated by transfer printing*, Nature Photons, v. 5, 2011 (online)
[3] Kim, Sangwook, et al. *Effects of Hf Incorporation on Negative Bias-Illumination Stress Stability in HIZO Thin-Film Transistors*, Japanese Journal of Applied Physics, v. 52, 2013 (online)
[4] Kim, Hyun-Sook, et al. *High Performance and High Stability Oxide Semiconductor Thin-Film Transistors (TFTs) for Next-Generation Flat-Panel Displays*, 2011 (online)
[5] Chesterfield, Reid, et al. *Solution-Coating Technology for AMOLED Displays*, Information Display, v. 1, 2011 (online)

---

* The authors of [8] named their TFT after M. Corbino, who designed a disk-shaped magneto-resistance device made up of concentric ring contacts back in 1911.

[6] Shin, Hwang, Yu, and Sung. *Design of Organic TFT Pixel Electrode Circuit for Active-Matrix OLED Displays*, Journal of Computers, v. 3, n. 3, 2008 (online)
[7] Keskin, Seyma. *Conducting Metallopolymers with Tridentate Ligands*, PhD dissertation, University of Texas at Austin, May 2018 (online)
[8] Lee, Hojin, et al. *Asymmetric Electrical Properties of Corbino TFT and Concepts of its Application to Flat Panel Displays*, IEEE Transactions on Electronic Devices, v. 54, n. 4, 2007 (online)

Chapter 15

Invisibility

15.0. Introduction

In this chapter invisibility technology is explored. Though some detail will be given, this is not an exhaustive mathematical overview and the reader is urged to consult the references.

The science of evaluating small particles in a host medium started out as models of atoms, and so the small particles were basically spherical. For this reason, inclusion models for metamaterials are dominated by arrays of small spheres. From a manufacturing viewpoint, small spheres are not hard to make, even on the nanometer scale, so that one strong candidate for an invisibility metamaterial is something made out of multiple layers of small spheres in a cubic face-centered or hexagonal arrangement.

Along these lines this paper will explore the technology of Liu, Zhan, and Chen [1]. Their "super absorber" is only a super absorber because of an exact choice of the material for the spheres; the material of the host medium; the sizes of the elements; and the ordering of layers. An earlier paper by Zhan *et al* [2] uses the same array of spheres but different materials and sizes, and in that case the material is an "anomalous transmitter". This paper will attempt to explain why there's such a difference between what seem to be identical designs; and the ramifications this kind of specific behavior has for the design of an invisibility cloak.

15.1. Physics Lore

There's a trove of lore behind the mathematics of invisibility that, amazingly for such a disparate set of tools, people mostly agree on. The theory of atoms as it comes down to us today goes back to the 1890s, and this long history informs the much newer field of metamaterial science.

One discriminates between *dynamic* and *static* equations for material properties, related of course to the electromagnetic fields impinging on and traveling through the atomic structure. The magnetic field determines a dynamic behavior; whereas a dielectric (non-conducting material) can be described as a static regime. In between is the *quasi-static* regime. For example, a very thin conducting layer on top of a dielectric layer might be modeled as a quasi-static regime, since for the most part the behavior of the dielectric dominates; when the magnetic field plays a secondary role to the electric, the quasi-static regime describes a dielectric. The dielectric is exposed to a slowly varying external electric field E. The same equations apply both outside and inside of the material. The magnetic field, if any, has a higher order of smallness than the electric field; and so the quasi-static limit for dielectrics corresponds to the *static limit.* In general, with the quasi- or normally static limit, the permittivity is almost always frequency dependent.

This poses a problem since invisibility in open air must be broadband: it must bend the entire visible light spectrum. Various researchers have come up with mathematical simulations or working prototypes of cloaks that are invisible to a specific single frequency, and it may be invisible when looked at with a lab instrument tuned to that frequency, but these designs won't work for the 3D display cube. Instead, a broadband solution that requires no laser pumping or exotic magnetics is needed.

As already mentioned, inclusions in the shape of spheres are the best studied case when it comes to mathematical theory. The size and material parameters of the nanosphere are factored into the *polarizability*, which relates the dipole moment of a sphere with the electric field acting on that sphere, which, in an array of spheres, is produced by all the other dipoles and by the field external to the system. [3]

The simplest cloak is spherical, but the 3D display device uses cylinders. The cylinder as a geometry is minimally more complicated than the sphere. Insight comes from the evaluation of a cylinder as an inclusion shape (rather than as a macroscopic shape). If the cylinder is compared to a spheroid, the main deviation in its response at the microscopic level comes from the field singularities of the wedges in the top and bottom faces of the cylinder, which a spheroid lacks. [4] These singularities would almost certainly lead to absorption losses in the material due to noise. However the macroscopic cylinder used in the display cube only consists of the cylinder walls, not the endcaps, and so the light response is much simplified. The response is dependent on the direction of the electric field. Unlike the sphere, the cylinder requires a geometric parameter to describe its shape: the length-to-diameter ratio. This extra information complicates the polarizability functions: one in the axial (sometimes called *longitudinal*) direction (z) and one in the *transverse* direction (t).

For the macroscopic cylinder, polarizability is also a factor in terms of the light that reaches the surface of the cylinder from the outside. But most of the difference is simply that macroscopic cylinders use a different coordinate system than spheres.

15.2. Tunneling

According to the author of [5], various resonant modes between complementary metamaterial slabs may produce wave tunneling, making the whole region of space occupied by the two materials invisible. This involves energy transmission via *surface plasmons*: a plasmon can be understood as a kind of electron or photon cloud that follows the contours of the metamaterial nanostructure. It's a *near field* phenomenon.

The interplay between metal and dielectric, the first conducting and the second non-conducting, is the key to the resonant modes that develop. A perfect electric conductor (PEC) is the idealization of a good conductor: the permittivity is so high that its inverse approaches zero – and as the inverse approaches zero so too does the non-reciprocal permeability. Analogously, a perfect magnetic conductor (PMC) is the dual of a PEC: as the inverse of the permeability approaches zero (meaning that the permeability becomes infinite) so too does the non-reciprocal permittivity approach zero. Of course a perfect *anything* is impossible because of signal losses due to material absorption, and due to dispersion during propagation.

The PEC core of a cloak isn't completely divorced from the material that surrounds it: it reaches out into that material via *delta* functions. [6] A delta function is attached to permittivity, or alternately to permeability, and pertains to the instantaneous response of the cloaking system when exposed to frequencies much lower than those at the microscopic scale. The reason metamaterials exhibit anomalous behavior is because their grid frequency is much higher than the impinging signal frequency.

A perfectly nonscattering cloak can't absorb power and thus the material of the cloak must be lossless. Just as there can be delta functions at the boundary between a material and its PEC core; so too there can be delta functions at the boundary between a material and the free space that surrounds it. Maxwell's equations imply that the *tangential* components of electric field E and magnetic field H are continuous across the outer surface of the cloak, unless, that is, there are delta functions in the fields at the outer boundary. Also, Maxwell's equations require that delta functions are produced in the *normal* components of E and H by any sources inside a cavity. On the other hand, the tangential-field boundary conditions at the outer surface of the cloak; and the normal-field boundary conditions at the inner surface, should be compatible with zero total fields inside the interior cavity of the cloak. [7]

Spherical cloaks with these boundary conditions have E and H fields that are continuously zero across their inner surfaces; but cylindrical cloaks have delta functions describing polarization densities at their inner surfaces. This is one aspect of the slightly more complicated geometry and behavior of the cylinder versus the sphere. To make matters worse, the authors of [7] have some bad news: the causality-energy conditions of the Maxwell equations imply that even single-frequency cloaking in the quasi-static frequency range of the individual dipole scatterers comprising the material or metamaterial is not realizable except for magnetostatic cloaking – that is, for a quasi-static system pushed into resonance via the magnetic field, which takes precedence over the electric.

Unfortunately, the resonant nature of most metamaterial designs to date leads to significant frequency dispersion, resulting in a narrow bandwidth for the desired response, and a large material absorption factor – the exact opposite of the desired non-scattering behavior of a tunneling material. However, advances have been made in this arena, as will be discussed in sections 5, 6, and 7.

15.3. Applications

Experiments over the last ten years with arrays of nanoparticles were inspired by earlier experiments with arrays of holes in a material. One such experiment involved an array of holes in a metal film resting on a dielectric substrate. As described in [8], surface electromagnetic resonant modes between the dielectric and film can be thought of as surface plasmon "molecular levels" arising from nonequal surface plasmon "atoms" – something like the quantum shift in energy which produces light. These surface plasmon molecular states are coupled to radiation modes, and extraordinary optical transmission (EOT) occurs via these modes. Transmission due to resonance depends on plasmon coupling and on the metal thickness. Transmission due to radiation, on the other hand, is tied to the material surface itself. All of this contributes to the tunneling effect through the material. Resonant peaks in the optical wavelengths only occur (so far) in the noble metals: silver and gold. All other metals except possibly aluminum are too resistive in the optical regime.

One experimenter in China (Zhan) has worked with nanospheres. Two of his papers, published some seven years apart, deal with this experimental technology. The first paper, published in 2006, is the background for the second. And yet, in the first paper, the array of gold-covered nanospheres acts as a transmitter; and in the second, the same array acts as an absorber. Both are active in the near-infrared rather than the optical frequencies, and yet both experiments are relevant to optical applications.

In the first experiment [2], the transmission effect is attributed to surface plasmon polaritons (SPPs). The transmission is enhanced by the diffractive coupling of evanescent SPPs to the metal film, which leads to a strong concentration of light at the metal surface, which in turn weakly tunnels through the corrugations in the metal and re-radiates by an inverse process on the exit side. The transmission results were normalized by Zhan *et al* to the transmittance of a pure quartz substrate for a given infrared spot size (800 micrometers). The main resonance had a transmittance of 23 percent, about 2.5 times the projected area of the gaps between the nanospheres. In other words, the transmission was 2.5 times what could be attributed to direct transmission through the substrate itself.

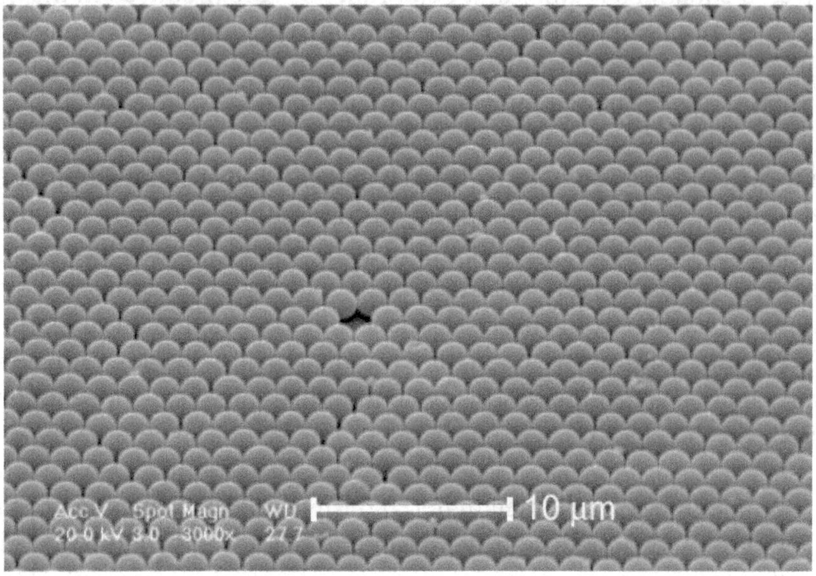

Figure 1. Scanning electron microscope image of the corrugated gold film deposited on silica spheres on a quartz substrate. The deposition is a "self-assembly" process; notice the defect. (Courtesy [2])

Zhan *et al* go through an extensive analysis of all of the factors in the construction, to try to determine what exactly is responsible for the extraordinary transmission. Their analysis is quite involved, but in essence the transmission peak depends on the thickness of the gold deposition, which takes the form of connected half-shells on top of the silica spheres.

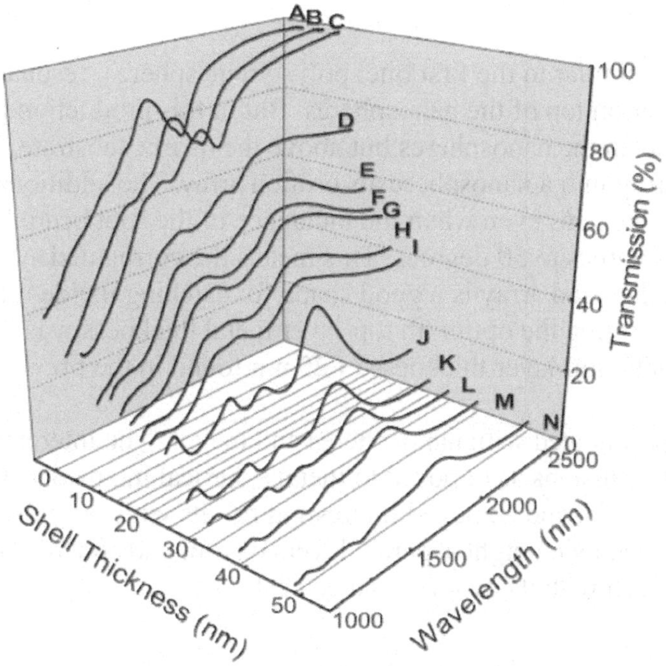

Figure 2. Transmission spectra of samples of different thicknesses of gold metal deposited on nanospheres. (Courtesy [2])

In the figure above, when the thickness of the metal layer is less than 14 nm (curve E), the metal film has no interconnections and is basically a collection of isolated gold caps on the nanospheres. However at 14 nm the gold makes contact with the substrate and photonic modes are activated in the dielectric crystal spheres. When the layer thickness is increased to about 25 nm, a conducting metal network is created on top of the colloidal crystal network of silica spheres; an abrupt decrease in the transmittance in the long wavelength limit is seen, combined with a sharp transmission peak with a high intensity: compare curves I and J above.

The transmission effect is tied to the thickness of the metallic conducting layer that covers the nanospheres. The authors also tried substituting an array of polystyrene spheres for the silica spheres, of the same size, and the results were the same, except that the transmission peaks were redshifted by about 200 nanometers. This is because polystyrene has a higher refractive index than silica (1.59 versus 1.45), which impacts the speed of light and alters the wavelength location. The higher index causes light to take longer to penetrate the colloidal layer. The same shifting effect can be obtained by making the sphere size (and thus the array size) smaller – and for the same reason. If light has a shorter distance to travel, the transmission spectra will be blueshifted. Whether red- or blue-shifted, the spectrum isn't compressed. The peaks seem to have the same spacing even with the smaller (or larger) array. Obviously, at some point the shift will start to become a compression (in the blue direction); or an expansion (in the red direction). For example, the same array exposed to optical waves would certainly have a completely different response.

Zhan used an infrared spectrometer to make the measurements; this instrument impinges an infrared beam on the sample and measures the response as the impinging infrared frequency is changed. Within the limited bandwidth measured (1000 to 2600 nanometers) the signal shift occurred when: a) the refractive index changed; or b) when the sphere size changed. But the entire response changed when changing the thickness of the gold layer because of the different states of contact between gold caps and between metal and dielectric. (Refer again to Figure 2.) The depth geometry and not the array spacing therefore drives the signal response.

Zhan's second paper is very similar to the first one: polystyrene spheres, resonance in the near infrared, and the deposition of a gold layer on top of the nanospheres. But in this production [1] Zhan *et al* added a reflecting layer of gold beneath the nanospheres but above the quartz substrate. This one addition turned the nanosphere transmission array into a nanosphere absorption array. The additional revelation in this second paper is that the absorption happens even when the incidence of the spot beam by which these effects are measured is varied from 0 degrees to 50 degrees. This means the corrugated system of connected gold half-shells on top of the colloidal crystal array is a good shape for catching incident light rays at almost any angle. Interestingly, for absorption the optimum top-layer metal thickness was 8 nanometers, considerably less than the minimum conduction-layer thickness of 25 nm found in the previous paper.

Other researchers have experimented with nanospheres and metals. The interesting results seem to center around dielectric spherical inclusions as opposed to metal spherical inclusions. One researcher [9] used a silver metal film rather than gold, and achieved interesting results with extraordinary transmission when doubling, quadrupling, or even using eight separate layers of film and spheres. The effect is to split the resonance into sub-bands, each with its own resonance peak.

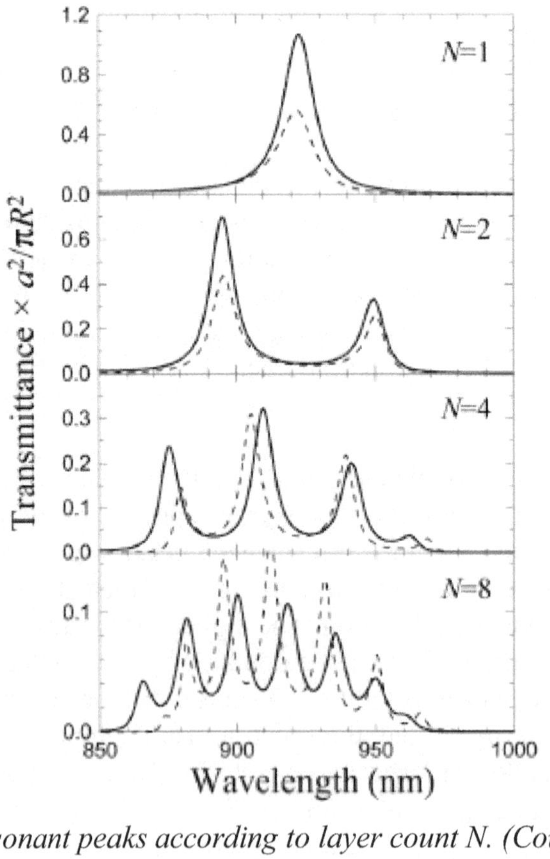

Figure 3. Resonant peaks according to layer count N. (Courtesy [9])

In terms of multiple layers, this researcher kept the distance between layers close enough to encourage the transfer of surface plasmons (the separation between planes was only 5 nm). On the other hand, the cell size within a plane was 250 nm, and the nanosphere diameter 150 nm, so that each sphere was separated by 100 nm – to prevent interaction between spheres within a single plane. This scheme could well be applied to the outer layer of a cylindrical cloak. Notice in the charts above that even when the spheres almost touch each other in the stacked direction, the transmission coefficient drops dramatically as the layers increase.

For this reason, the author proposed changing over to an active scenario with laser pumping or an applied electric field. The cube can ill afford an active system because it's already burning power to drive the LEDs. Putting cloak batteries and cloak drivers on each spacer level is simply not an option although a passive electrical charge isn't ruled out.

Another researcher working with layers [10] noted that a smaller distance between paired spheres (in two different layers) resulted in a loss of accuracy of his single-dipole mathematical approximation. He thought that when the nanospheres are so close together, higher order spherical harmonics or other full-wave numerical models should be used. But there's a tradeoff: the higher the harmonic, such as with quadrupoles or octopoles, the lower the achieved power. Dipoles attain the most power.

In the general case, the polarizability equation takes into account the size and material parameters of the nanospheres. Polarizability relates the dipole moment of the sphere with the electric field acting on that sphere, which is produced by all the other dipoles and by the field external to the system. [11] The polarization at any particular point of the nanosphere is influenced by the field at other points on the sphere. For inclusion shapes other than a sphere, the electric and magnetic responses will be strongly dependent on the direction of the propagating wave and often also on its polarization. [12]

15.4. Effective Medium Theory (EMT)

How good is good enough when modeling atoms? The two approaches to modeling metamaterials, microscopic and macroscopic, both have their benefits. In the *macroscopic* approach, since each element is much smaller than the wavelength of the frequency of operation, individual elements are considered as a homogeneous, effective medium with an electromagnetic response defined by an effective permittivity and permeability.

To investigate interparticle coupling at the microscopic level, one can resort to the single dipole per particle approximation (SDA) – whereby each particle is modeled with just one electric dipole. In this case dipole interactions among particles are modeled correctly, and the SDA is accurate for particles that are small and well separated, in which case the field of each particle at any of its nearest neighbors is well represented by the dipole field. Each particle requires a "dipole polarizability" expression. [13]

Tightly bound into the SDA equations for an atom's resonant angular frequencies is the Clausius-Mossotti formula, which relates a macroscopic property (e.g. the dielectric constant) to a microscopic property (e.g. the molecular polarizability). The relation was used by Albert Einstein in 1910 in his treatment of critical opalescence – e.g. enhanced light scattering near the critical point of a fluid due to large density fluctuations. Further polarizability extensions are found in the Maxwell-Garnett theory, which is one of the foundation stones of EMT. [6]

According to [14], the *effective medium limit* is the point at which the periodic structure of the material appears homogeneous to waves propagating inside it, and it is in this regime of homogeneity that one can attribute to the material effective properties such as permittivity, permeability, refractive index, and wave impedance, among other things. Because electric and magnetic surface charges and currents are zero in a lossless material, Maxwell's equations imply that the tangential components of E and H (and their normal components) are continuous across the outer surface S of the material, unless there are delta functions in the fields at S. This is mathematically important.

At the root of EMT is approximation and averaging. Averaging can be defined for EM fields and polarization so that the usual boundary conditions for a homogenized composite medium are satisfied: in other words, the tangential components of the averaged fields E and H are kept continuous across the boundary. The continuity can be preserved by introducing a very thin Drude transition layer across which the permittivity and permeability of the homogenized medium vary from the material bulk values and take on the values in the surrounding space. The effect of a finite-thickness boundary of a composite medium becomes important at high frequencies and when the composite layer is thin, i.e. it extends only a few unit cells across. [12]

For the display cube, one aims for a passive medium because of its simplicity, but in reality an applied voltage and current, or magnetic field, might be involved (see Section 7). According to [12] all passive materials are causal, and in causal media frequency dispersion leads to signal loss. EMT is most successful when the material being modeled only suffers from negligible losses – which means that absorption must be negligible. Luckily, absorption can be ignored when the operational frequency is far from the resonant frequencies of the inclusions.

An active medium stores energy: to create fields in a material, work must be done to polarize the medium, which means that in the absence of losses, more energy must be stored in the material than in the surrounding empty space (also called *free space*). If a material has considerable losses near the frequency of interest, it's not possible to define the stored energy density in a general way. More precisely, it's not possible to express stored energy in terms of the material permittivity and permeability functions. Knowledge about the material microstructure becomes necessary, and the problem turns into one of discovering reactive energy densities.

The accuracy of a polarizability expression, whether active or passive, is impacted by the material lossiness, which in turn is driven by the array spacing, the particle size, and other factors. In the static regime, a lossless nanosphere (which by definition has imaginary material *emittivity* = 0), to satisfy the law of energy conservation, should have an absorption efficiency Q that's negative. This would imply that the energy is being created within the nanoparticle, which is counter-intuitive. For this reason, for small spheres and small losses, "radiation damping" is taken into account. With this factor in place, the energy conservation law is exactly satisfied and the efficiencies Q for extinction, scattering, and absorption are all zero. [10] This is the ideal state for extraordinary transmission and extraordinary reflection. Of course, this is theory: the practical realization of a lossless medium is another matter altogether.

15.5. Negative Index Materials

A negative index material (NIM) is characterized by negative refraction. For this to happen the effective permittivity and permeability must both be negative over the same frequency band. When the real parts of permittivity and permeability possess the same sign, electromagnetic waves can propagate through the NIM. If the signs are opposite, waves can't propagate unless the incident wave is "evanescent". [15]

Of particular interest for the invisibility cloak application is the behavior of an NIM slab (or "left-handed material") sandwiched between two "right-handed" materials. Negative refraction conducts the incident waves and focuses them in the trailing RHM: in other words, the NIM acts as a lens. But unlike a normal lens with a single focus, the NIM can focus multiple point sources – anywhere in the plane of the NIM slab. There is no unique optical axis. Parallel light rays (from infinity) can't be focused; rather, any number of sources at various displacements will be refocused in the trailing RHM. The following diagram shows how slab thickness *d* is equal to source distance *s1* plus external focal length *f2*.

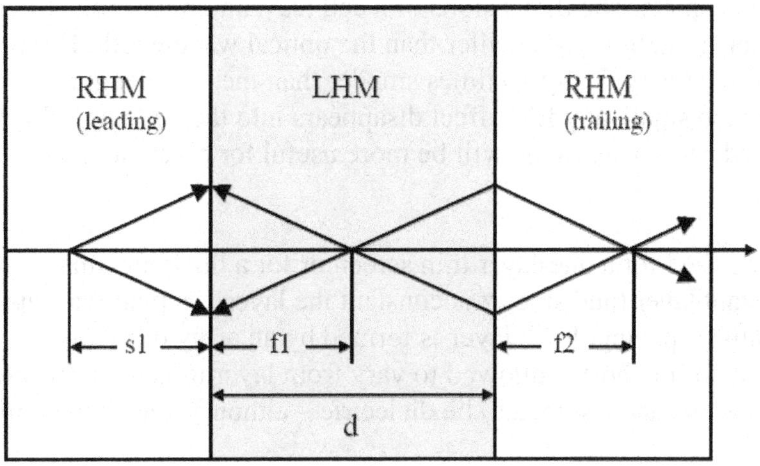

Figure 4. Negative refraction at work. Light travels left to right.

As the source in the leading RHM is moved towards the first interface, the external focus in the trailing RHM moves away. Contrariwise, if *s1* increases, *f2* will decrease. This kind of negative-index behavior might be quite useful for entraining incident light for an invisibility cloak transformation. Of particular interest is an additional fact: the NIM incorporates the principle of phase compensation: the phase *lag* incurred in the RHM regions between the source and external focus is exactly compensated for by the phase *advance* incurred in the LHM slab. This means the propagating spectral component of the source is completely restored at the focus – a critical requirement for a good invisibility cloak. [14]

15.6. Transverse Modeling

Based on the four-layer cloak designed by Zhang and Wu [16] it seems the only hope of realizing a cylindrical solution is with layers. A good approach for modeling layers is effective medium theory (EMT). The basic concern is getting the best approximation possible so that the simulation is as accurate as possible.

Silveirnha and Fernandes [17] have gone a long way towards realizing that goal. They first acknowledge the general approach of describing very thin metamaterial screens, most of the time with only one layer of inclusions, as bulk materials characterized by an effective permittivity and permeability, and of retrieving the effective parameters from measured or simulated S-parameter (scattering) data. They call this the "classical homogenization approach." The classical approach may not be adequate when the metamaterial screens are very thin – the extreme case being a single layer.

The authors propose using a transverse averaged (TA)-field approach. Because the fabrication of multi-layer metamaterials is still a challenge, most prototypes have four or fewer layers. The material is manufactured using planar technology, as a sheet; mathematically there's no problem with treating the periodic structure as an infinite lattice. The fields inside of the slab are modeled as the superimposition of the average modal fields (E, H) of the infinite lattice, subject to the classical boundary conditions (being the continuity of the tangential components of E and H). The boundary conditions are enforced at the slab interfaces. If the slab is in the (x, y) plane, then the stacked direction for multiple layers is in z. The z-direction is the direction of the transverse field. This means that the TA approach is tailored for describing the effective parameters of multiple layers.

In natural materials the characteristic dimensions of a cell (say, an optical cell structure) is about 0.1 nanometers – three orders of magnitude smaller than the optical wavelength. But in typical microstructured materials the lattice constant is typically 10 times smaller than the wavelength. Still, it has been shown that if the lattice periods are too small, any EM effect disappears into the general field. Thus the inclusion size in the ten to one-hundred nanometer range will be more useful for practical purposes than the fine sizes found in nature.

The TA-approach can be used for a one-layer thin screen or for a thick metamaterial slab with multiple layers. For each z-constant layer (and since z is constant the layer is a plane) a transverse "unit cell" is defined based on the lattice spacing. Each layer is formed by an array of inclusions embedded in a host medium. The permittivity of the host is allowed to vary from layer to layer. (Permittivity becomes a function of z.) The inclusions are assumed to be dielectric – although they can be metallic and conducting too.

As usual, polarization enters the picture and complicates things. However the complications can be sidestepped provided that the effective parameters of the bulk metamaterials are known in advance. This may seem counter-intuitive, but is in fact a valid approach, even if it means starting with assumed values and "working backwards" by iterating over scattering solutions until finding one that works.

The z-based homogenization approach satisifies the condition of tangential components being continuous for microscopic fields (E, B). The difference between transverse averaging and the "classical" approach is that TA satisfies this condition at every z = constant plane containing dielectric inclusions. However the transverse-averaged fields can be discontinuous when the inclusions are "perfect electric conductors" which are parallel to the layer interface and which induce a surface current. The authors treat this condition by modeling the cell with a permittivity equal to minus infinity. But if the wires or inclusions are parallel to z, a singularity in the current density J results, and an additional boundary condition must be imposed.

The tangential component of magnetic field H is discontinuous across the interface of a perfect electric conductor by the amount of surface current density J. In other words, the density of current J doesn't vanish at the interface. Basically the current density J depends on the incident radiation on the surface. It's perfectly feasible for incident electric and magnetic fields to interchange their roles, so whether time variation of the magnetic field affects the electric field or *vice versa* isn't really important. For expendiency, the TA theory assumes that there are no magnetic inclusions, and that magnetic effects, if they exist, arise due to the circulation of electric currents. One more assumption is that the inclusions don't intersect the z = constant plane of interest, which is a natural assumption: different inclusion geometries are bounded by their own layers. However, inclusions can *abut* the z = constant plane without penetrating it.

The TA method treats the fields inside the slab as a superimposition of the transverse-averaged electromagnetic (EM) modes, not as a superimposition of cell-averaged EM modes. In this way, the important variations of the fields along the z-direction are properly taken into account, as opposed to the "classical" method wherein the variations are lost. In particular, the polarization properties of the TA-fields may vary with z. Once the TA-fields for the infinite lattice have been computed, the scattering problem can be easily solved.

The reader can refer to [17] for mathematical details. As with a homogenized medium, the TA fields can be decomposed into transverse-electric and transverse-magnetic modes with relation to z. In most cases the polarization of the TA field is the same as the polarization of the bulk medium fields. The difference is that the averaging for the (E, B) modal fields in the general EMT approach is uniform and independent of z; while the impedances seen by the TA fields may vary strongly with z. This is a manifestation of the granularity and discreteness of the periodic structure. Another difference is that the transverse impedance seen by a wave that propagates along the positive z direction is, in general, different from the impedance seen by the mode that propagates in the opposite direction. This is good information to have, since for a cylindrical *transformation geometry* the light rays will enter the media on one side, be bent by the properties of the materials, and exit the media on the other side, and thus travel in both directions through the media relative to z.

One more detail is that the propagation constant *k* for the transverse-electric mode is a function of frequency, of the angle of incidence, and of the effective parameters of the bulk medium. This means that all of these variables will affect how the TE mode is excited inside of the metamaterial slab. The constant *k* determines the reflection and transmission coefficients for transverse propagation. This means, in turn, that a complete model over many angles of incidence will require solving the equations for each of those angles and finding, within the constraint of no more than four layers (say), the one solution that works best for those four layers and for the range of desired incidences. Zhang and Wu [16] used a *genetic optimization* algorithm to solve a similar problem. In that case they used the discrete values of permittivity and permeability generated by the general transformation equations, and, limiting the design to four layers, found the optimized solution for those four layers, thus minimizing the scattering cross-section of the cylinder under those constraints.

One more note: while TE modes are measured in the "transverse" direction, *transverse-magnetic* (TM) modes are measured in the *longitudinal* direction. Some researchers seem to ignore the longitudinal direction entirely; but it can't be ignored if modeling negative-index materials since some of the left-handed behavior is indeed "magnetic". (See Figure 6 in the next section for an example layout.) Towards that end, the authors of [17] tried their math on an array of infinite-length metallic rods in the (x, y) plane. An array of *nanorods* (also called *nanowires*) is an approximation of an ENG or epsilon-negative material, epsilon referring to permittivity, e.g. to the dielectric constant. For the rod geometry, the real part of the permittivity is negative. (This is why the PEC nanowires are modeled with a negative infinite permittivity.)

One sidelight on the way the authors of [17] present their results. The graphs are drawn with respect to the normalized frequency (ω / c) times the separating distance between rods, *a*. Thus the graphs are valid for any material. At ultraviolet frequencies most metallic materials lose their conducting properties and become "transparent". For example, silver becomes transparent at a wavelength of 1386 angstroms, in the ultraviolet. If (ω / c) is the reciprocal (more or less) of the wavelength at which silver becomes transparent, or (1 / 1386), then the behavior at (ω / c) * *a* would hold at normalized value 1 if the distance *a* is 1386 angstroms (138 nanometers). Obviously negative-index behavior for some wavelength can only be verified by actually building the array. (See also [18].)

When a metal becomes transparent it behaves like a dielectric. It transmits incident radiation and reflects some too. The transparency frequency is also called the *plasma frequency*. The ENG signature seems to be most prevalent at plasma frequencies – but the plasma frequency for a metamaterial can be lower than for a metal like silver. At optical frequencies a metal like silver doesn't absorb much, and reflects most frequencies. Silver has a reflection coefficient of nearly 1 across the entire visible spectrum and so has a whitish color; gold selectively reflects through its lattice ions at the red end, giving the well-known gold color.

In general, silver or gold reflect most light at optical frequencies and only absorb about 1 percent of the incident energy. That's why the gold backing of the Zhan super-absorber is so efficient: the energy is reflected by the backplane and absorbed by whatever's in front of the backplane.

As will be discussed next, the surface current distribution J on a waveguide is always in motion and associated with transverse electric modes. It's important to realize that a waveguide is so complicated that the values for (current) I and (voltage) V depend on where in the waveguide one attempts to compute them. The "impedance" which is defined as the ratio V / I is therefore not a clearly definable or unique quantity. But under controlled circumstances impedance has definite measure.

15.7. Layered Structures and Modal Properties

Negative-index materials can be subclassified as ENG (epsilon-negative), MNG (mu-negative, referring to permeability), or DNG (double negative, referring to both parameters). The DNG signature can only be achieved with metamaterials; and the MNG signature with negative permeability and positive permittivity applies to a sub-class of materials that are *gyrotropic*.

One more acronym is DPS: a double-positive system associated with common dielectrics. DPS slabs are known for surface polaritons, surface plasmons, or surface evanescent waves. These surface waves tend to form on metal-dielectric interfaces, where the plasmon resonances form "slow modes" that travel at less than the speed of light. [19] This raises the refractive index at the boundary and can contribute to total internal reflection. The E and H fields are evanescent, and decay with distance away from the boundary surface. The Lorentz Dispersive Model is typically used to model dispersion of the propagating signal or wave packet along the surface. [23] In a vacuum all frequencies travel at the speed of light; but in the presence of a material the refractive index varies with the frequency of the propagating signal. The attenuation constant for dispersion also varies with the signal's frequency and polarization.

Because of dispersion, all materials to one degree or another are *lossy*. But losses decrease as the field intensity increases, and this happens at a DNG – DPS boundary. Intensity at the interface is strong, and the attenuation of the signal with increasing distance from this critical boundary is dramatic. This means that thick materials add nothing to the behavior, and that thin layers are fully functional for negative-index applications. This insight will play a role in the cloak design for the display cube.

Negative refraction is described by the triplet (E, H, k), where E and H are the EM fields as usual, and k is a propagation constant as usual. Vector k is reversed in DNG media. A DNG slab can be made "very thin," thus increasing the confinement of the field and its fundamental mode. Note that a DNG material can only be created when the resonances for permittivity and permeability *at the same frequency* are close to each other. Resonant modes that are completely divorced from each other will *not* result in negative refraction. In general resonances arise from fundamental modes, and these modes can change depending on how dense the inner (DNG) material is compared to the outer materials. [23]

By properly placing a DPS slab near a DNG slab, the different handedness of the media causes a contradirectional power flow, so that a "backward coupler" can be achieved. To frustrate total internal reflection a jagged or bumpy surface can act as a coupler, which is how high-speed optical *grating couplers* work. In essence, the Zhan bumpy surface of gold on dielectric can act as a fancy grating coupler from free space into a negative-index medium waveguide, and from the waveguide back to free space, with the key transfer process being plasmonic. Backward coupling could be a major win for a broadband cloak, because light has to both couple into a waveguide and exit out of the waveguide at the end of the trip. The exiting of lightwaves could be aided by backward coupling.

Along the lines of [23] the author of [20] writes about an *indefinite medium* (IM) as a class of metamaterials for which not all of the principal elements of the permittivity and/or permeability tensors have the same sign. Again, the unique dispersion properties of such materials permits extreme optical states and allows broadband negative refraction (the most well-known NIM characteristic) at indefinite / normal materials interfaces.

Yao [20] shows how the tensor elements for a uniaxial indefinite material can have different signs. Yao like other authors uses an array of nanorods to illustrate his case; k_P is the propagation vector along the length of the rods (in z). Meanwhile, k_V is the vector sum of the x and y components and thus represents propagation perpendicular to the rod axis in z. For a fixed frequency these two propagation vectors can be plotted as an isofrequency contour in the (k_V , k_P) plane. As a result of the negative sign of the z-component of the permittivity, the result is a hyperbola. In fact, the curve is completely divorced from the frequency contour of free space (air), which is a small circle at the origin. In general all dielectrics take the form of a closed curve: an ellipse or circle. This is shown below.

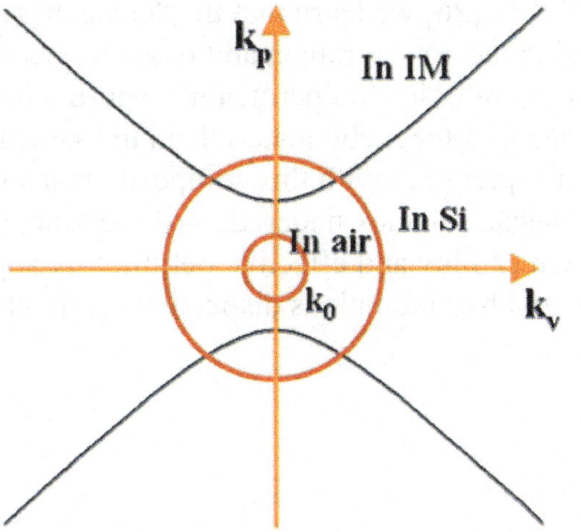

Figure 5. Fixed-frequency isocontour for wave propagation. (Courtesy [20])

Another way to describe this is as a *dispersion* relationship. The wave vectors in the radial and tangential directions, along with electric permittivities in those same directions, are equal to the angular frequency of the electromagnetic wave. If the medium has an isotropic response the electric permittivity is the same in all directions and corresponds to the circle in Figure 5. Circular dispersion is undesirable because wave vectors are limited to the refractive index of the medium. On the other hand, materials with anisotropic electrical responses can support the propagation of large wave vectors. If the radial permittivity is negative and the tangential positive, the dispersion relation now represents the hyperbola, as above. With a hyperbolic dispersion theoretically there is no limit to the maximum wave vector in the tangential direction that can be supported by the material. [27]

The indefinite material is modeled with silver nanowires embedded in corundum (synthetic Al_2O_3). Presumably the host material is transparent, as with sapphire. This nanowire system allows propagation of waves with very large momentum, and introduces a strong mode mismatch between the underlying metamaterial and the surrounding medium (e.g. free space) which results in total internal reflection. Coupling from plane waves in free space to an indefinite (resonant) cavity is limited due to the large mismatch of wave vectors (as in Figure 5: the air contour and IM contour don't overlap). However, evanescence through the TIR layer significantly improves the coupling efficiency.

Interestingly, cavity modes correspond to different wavelength energies. Modes (2,1,1), (2,1,2), and (2,1,3) resonate at vacuum wavelengths of 1500, 918, and 750 nm respectively. This is intriguing because three points in the optical bandwidth can be chosen for red, green, and blue frequencies. In other words, a cavity can have three operating modes depending on the wavelengths. The mode mismatches between the cavity modes and those of free space result in very low radiation losses, since the high-k components can't couple from bulk indefinite material into air.

Moving on, the mu-negative (MNG) signature relates to *gyrotropic* materials and seems to be the least-well understood. In this regime a *nonreciprocal* waveguide is defined by the interaction between (E, H) and (J, M), being electric power density J and magnetic power density M, related to a radiating source. [24] In short, the interaction of a source *a* in an EM field *b* produced by a source *b* is not reciprocal. The surface will not support the source *b* acting on the field *a* produced by the source *a*. [25] The most common nonreciprocal media are magneto-optical materials (bismuth-iron-garnet and indium-antimony at optical frequencies); and gyromagnetic materials (ferrites at microwave frequencies).*

This brings us to *gyrotropic nihility*. In [26] we learn that the "characteristic" frequencies for permittivity in magneto-optic media are confined to the optical range; and those for permeability to the microwave range. The characteristic frequencies for permittivity and permeability normally lie far from each other; but by combining together gyroelectric and gyromagnetic materials in the same host medium, one makes a *gyroelectromagnetic composite*. It is precisely with this composite that a negative index metamaterial can be realized. The problem with all negative-index materials is dispersion, but an "anomalous dispersion" can be achieved when the effective permeability and effective permittivity approach zero. This is called a *nihility state*. In this state the material has anomalous dispersion, or in other words, almost no dispersion at all.

---

* It should be noted that indium, bismuth, and antimony are native semi-metals that are mildly poisonous. The author of [25] seems to prefer yttrium-iron-garnet as a gyromagnetic material. The story with yttrium is this: In 1794 the Finnish chemist Gadolin found a "new earth" in a specimen of rare-earth pegmatite, which later yielded the metal to be known as yttrium. Yttrium is a purification of Gadolin's impure "yttria." The name came from the town of Ytterby in Sweden; the mineral was first found in a nearby quarry. Yttrium is a full-on metal though it's unclear how it's extracted from pegmatite, and what the "rare-earth" component of the pegmatite actually is. Scandium via thortveitite may provide the answer: it's also found in Ytterby, and the rare-earth mineral in the granite pegmatite it's found in is identified as uranium. Scandium is in fact a yttrium silicate and a byproduct of uranium ore processing. Thus yttrium is not only poisonous but radioactive and its use should probably be avoided. Recall the use of yttrium-barium-copper-oxide as a superconductor.

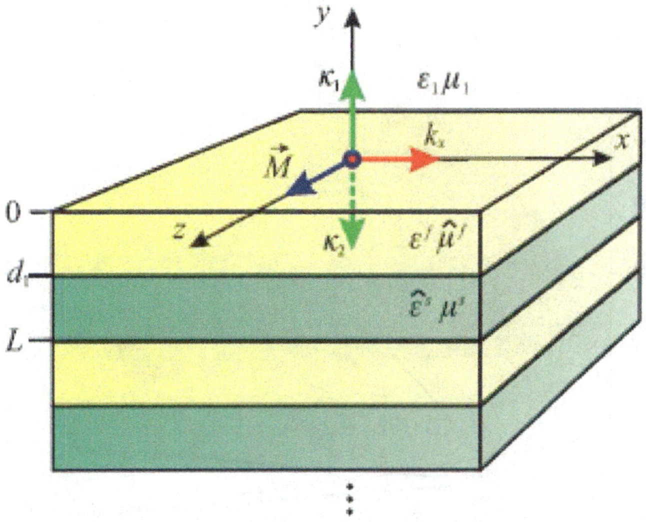

Figure 6. Very thin ferrite-semiconductor layers. (Courtesy [26])

Tuz proposes the geometry of Figure 6. The stack possesses a periodic structure that fills the infinite half-space y < 0. The outside space or free space takes up the half-space of y > 0. Each region or layer has its own permittivity and permeability. A magnetic field is applied along z, shown as M. The wave propagates along x, shown by k_x. In the cylindrical view of the cloak world, z is the height of the cylinder; x goes around the cylinder circumference; and y is the very-thin cloak thickness.

The attenuation of the signal along the y-axis, away from the interfaces, is shown by k_i, where *i* is the *n*th layer. The author emphasizes, quite rightly, that the individual layers are to be seen holistically via the application of effective medium theory. In this way, the finely-stratified multilayer system is approximated as a biaxial anisotropic (gyroelectromagnetic) uniform medium. The medium has two optical axes: in y and along z. Because a magnetic field is involved, polaritons are formed and lead to the gyrotropic nihility state in the bulk medium. To find this state the parameters of the stack must be put through an optimization procedure. The characteristic frequencies and magnetic field strength are fixed, while the other parameters are allowed to vary.

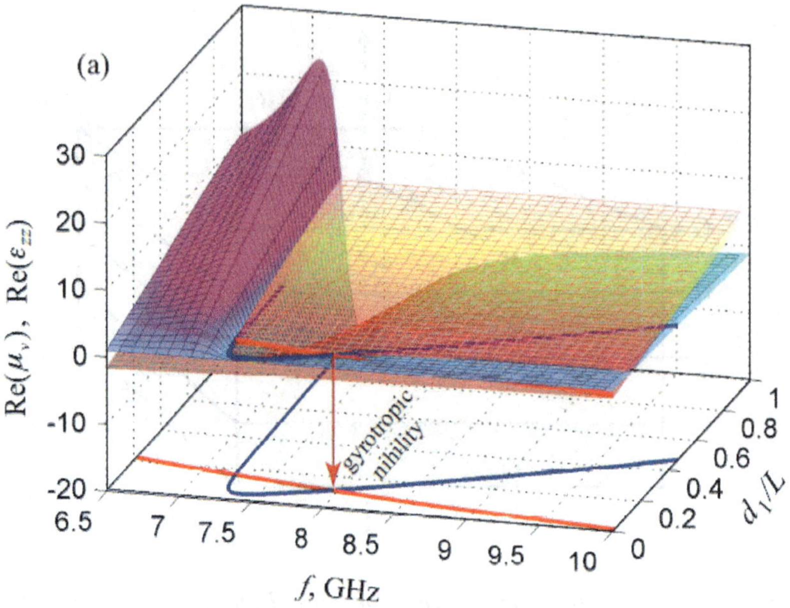

Figure 7. Results of the optimization. (Courtesy [26])

As shown in Figure 7, when the permeability and permittivity are both zero, the dispersionless state is achieved; the projection onto the groundplane shows that the optimum alternating (normalized) layer thicknesses are $\delta_1 = 0.117$ (for ferrite) and $\delta_2 = 0.883$ (for semiconductor).

All of this is critical for answering the question: how do you get ambient light into the metamaterial, get it transmitted, and have it emerge from the metamaterial without losses? Can light be routed via a curved waveguide around the circular core of the cloak? Normally this flow-behavior is locked into the parameters of the cloak and one doesn't think in terms of creating a waveguide. But in the end the parameters of an NIM waveguide as such, and of an invisibility cloak as such, may be equivalent.

At infrared and optical frequencies metals can't be characterized as good conductors. (As mentioned in Section 6 they are good reflectors, but they are not good conductors.) They are lossy solid-state plasmas, with a complex permittivity. Near the plasma frequency, which for most metals is in the ultraviolet, the absolute value of the permittivity divided by the free-space permittivity (ε_0) is about 1. However, at infrared and, in many cases, also at optical frequencies, most metals satisfy the inequalites $|\varepsilon| / \varepsilon_0 >> 1$, and $|\mathrm{Re}(\varepsilon)| >> |\mathrm{Im}(\varepsilon)|$. When these conditions hold, it's still possible to design a circuit model using shapes like split-ring resonators (SRRs). The ring will have capacitance and current, including displacement currents, and a ring-magnetic inductance can be defined for the magnetic flux across the ring.

However SRRs will not work at optical wavelengths as waveguides. When the rings get small, the frequency of resonance saturates to a fixed value based on inductance and capacitance, which doesn't depend on the ring size. Likewise, the loss factor saturates to a value based on resistance and inductance. This is the main limitation of scaling down SRR media to achieve a negative magnetic permeability at infrared and optical frequencies: the saturation of the frequency of resonance is directly tied to a decrease in magnetic response. [21]

Alu and Engheta [22] realized that arrays of closely spaced nanoparticles eventually resemble stacks of planar layers of dielectric and plasmonic materials which correspond to a simple planar geometry. When the electric field is properly polarized, these stacks of plasmonic and non-plasmonic materials may constitute effective, layered, negative-refractive, optical metamaterials in the form of left-handed nanotransmission lines. Unfortunately, the state of the art will only carry a signal about 180 nanometers – well short of the display-cube cloaked control-line half-circumference of, say, 100 micrometers. But the authors disclose a more interesting fact: in the limit at which the thickness of the inner (guiding) layer is small compared with the wavelength, the stacked structure still supports a guided mode. The wavenumber may become very high, and this corresponds to a "slow-mode" operation. The term *wavenumber* is just a name for the propagation constant k, e.g. 2π divided by the wavelength λ. Therefore a high wavenumber corresponds to a very small wavelength and thus a very high frequency. A slow mode equates to a high index of refraction.

The slow-wave (or slow-mode) behavior that happens in these cases is explained by the fact that in these subwavelength structures the field distributions are very much concentrated around the interfaces $\gamma = +/- d / 2$, d being the waveguide depth in the transverse direction. The depth is very thin. There are in fact a whole group of modes that can be identified in this quasi-static regime of thin layers. The transmission of light in these circumstances is affected by these modes; the phenomenon is similar to transmitting a signal via a subcarrier. In other words, light signals would be carried along the same way a television image is transmitted over a modulated subcarrier. A more picturesque analogy is people using a subway train. Light would be entrained by the refractive boundary, and if the metamaterial thicknesses and wavenumbers are correct, the light would be released without dispersion at an appropriate distance from the entry point. The hard part is getting the passenger on and off without losses.

The TIR of an indefinite material confines the signal, as mentioned in [20], but also blocks it. This could be overcome for coupling by using the evanescence associated with anomalous transmission, as in [2]. Evanescence on the boarding side could be accomplished by a *canonical* anomalous transmitter (CAT) via the Zhan dielectric-nanosphere metallic-array approach. On the departure side, a *reverse* anomalous transmitter (RAT) would provide backward coupling. Since this on and off behavior happens everywhere on the circumference of the cylinder, the on and off points would alternate at small radial intervals of the cloak.

15.8. Power

As yet, a passive solution, even with all of these tricks, won't generate the power needed to drive the signals the needed distances. Alu and Engheta point out that the permeability of the metamaterial is directly affected by the longitudinal component of the electric field. The nomenclature is tricky, but if the electric field is perpendicular to the cylinder surface and in the direction of the waveguide, as shown below, the permeability is basically associated with the magnetic field which is parallel to the cylinder surface.

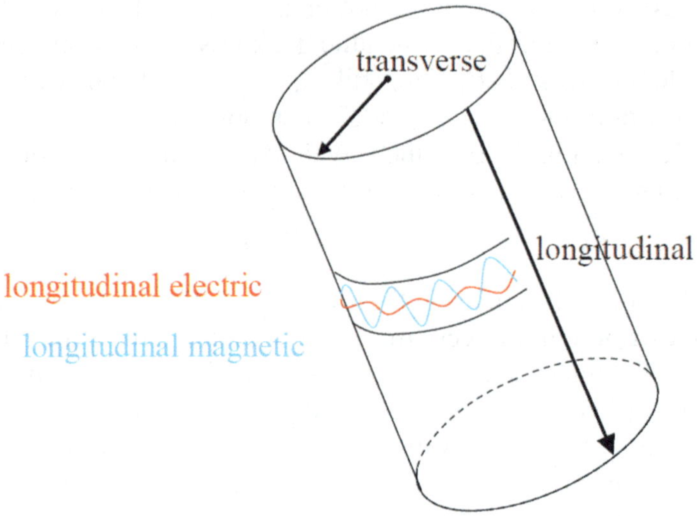

Figure 8. Cylindrical directions and waveguide nomenclature.

The EM fields are confined by the planes offset by depth(s) d. An artificial electric field traveling around the cylinder, or a magnetic field along its length (as in Figure 6), would enhance the permeability, the current density J, and thus the negative-index propagation. But discouragingly, the theoretical magnetic field used for [26] was rated at 2 kilogauss. The field was chosen to be strong enough to form a homogeneous saturated state for the ferrite as well as semiconductor sublayers. The lattice plasma frequency for the simulation was in the gigahertz range (Figure 7); compared to the thicknesses that would apply for the optical regime, the layers being solved for were three orders of magnitude thicker.

The cylindrical cloak for the display cube is very small – it has a diameter of about 80 mm. The scale of the invisibility application makes a difference: cloaking a jet plane is a very different problem than a cloak for a voxel-scale transmission conduit. The magnetic field will be very different for the two applications. For the very thin layers associated with an optical application over small distances, the magnetic saturation that drives the invisibility effect will correspond to the driving distances. The hope is that the magnetic field strength can be reduced to the milligauss range. The author of [26] ended his paper like this: "Experimental verification is still a challenging task. It involves solving an optimization problem to search for magnetic field strength, and corresponding thicknesses and the number of layers within a practical system. Nevertheless, we believe in its feasibility."

15.9. A Cylindrical Cloak

A preliminary design for a cylindrical cloak will now be given. The work of Zhan ([1], [2]) seems like a good start for the external interface: getting light into and out of the waveguides. Zhan's anomalous transmitter [2] can interface with each DPS / DNG port to get waves *out of* the guide. The transmitter can be reversed and thereby interface with an ENG / DNG configuration to get waves *into* the guide. Both configurations can be classified as pseudo-grating couplers.

Recall that all of the cloaks of the 3D display device are embedded in glass. Each spacer layer in the display cube has, at 128-cubed resolution, in the final design of Chapter 18, (4 x 8 x 8) + (4 x 8 x 8) = 512 holes for cloaks, each of which needs to be deposited, spin-coated, and/or etched with the appropriate materials at the appropriate thicknesses. Assuming that each hole containing each cloak can be spin-coated with thin layers, the actual cloak material will take up very little radial space. This is in contrast to most existing multilayer designs, which always show a cylinder composed of chunky metamaterial. The thin design is made possible by negative index materials.

NIM near-field focusing, though it won't image parallel rays, will transmit multiple light sources across a wide area. Ambient light needs to be captured at all areas of the outer waveguide surface, transmitted along the wave vector, and re-emitted. Since for the regular cube operation voltages are being supplied to the various color planes to drive the LEDs, these voltages could also be applied to the cloaks.

In this regime the light hitting the surface at angles close to the normal to the surface will be transmitted most strongly: as desired. Imagine a single cylindrical cloak. A viewer on one side of the cloak should see a second viewer on the other side without interruption – and *vice versa*. A lamp on a table to the first viewer's right-hand side will cast rays that hit the surface of the cloak normally about 90 degrees from the viewer's position – the light from the lamp will hit the cloak material at a glancing (oblique) angle relative to the viewing angle, and so the transmission will be very weak relative to the viewing position. All of this is desired behavior and is one way the cylindrical shape can work for the design.

Below, a layout as viewed from the side of the cylinder. The cube application has a *preferred orientation* because the principal light-travel direction is in the cylinder's radial direction. Four of the five cube sides which are viewable are viewed along this radial vector; only the top face has a downward or longitudinal view. Therefore the side view of Figure 9 shows the cylinder axis pointing in the horizontal direction. The cylinder as depicted in Figure 9 is resting on its side. Imagine that, underneath these access strips, the propagation of the waveguide runs vertically, in the radial direction.

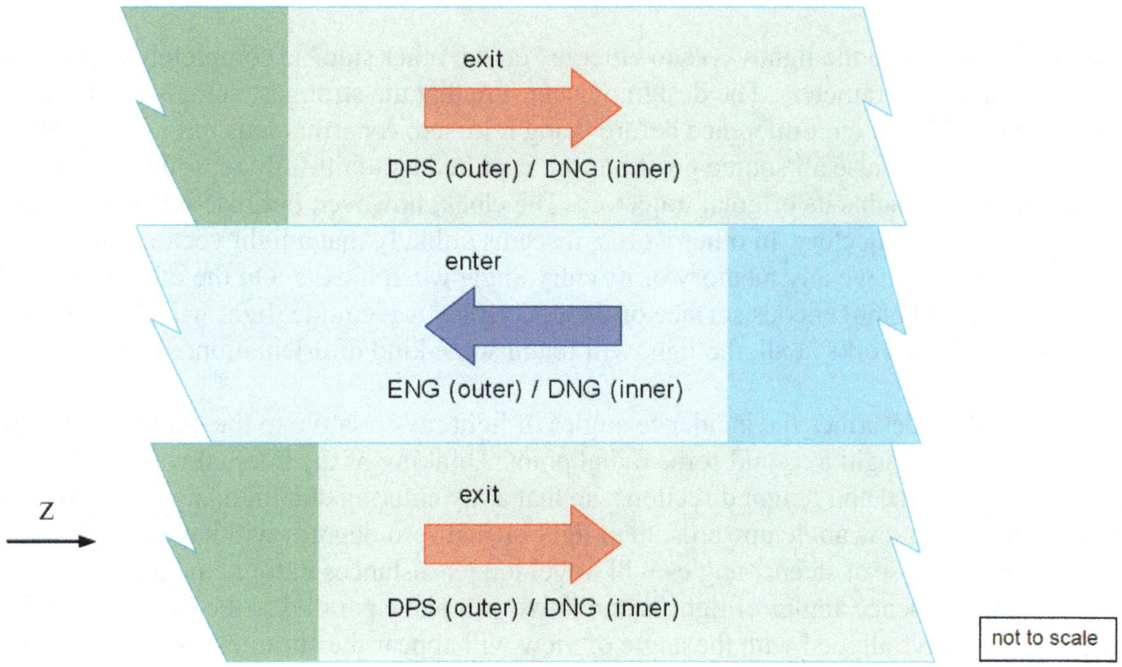

Figure 9. Side view of cloak showing simple waveguide access ports.

These alternating strips are spaced at the ten-micrometer scale, and so are indistinguishable to the naked eye. Light from any angle is captured and travels around the cylinder most strongly in the direction most closely aligned with the arrival vector. The next diagram shows a top-down view. At the micrometer scale, curvature of the cylindrical surface is negligible.

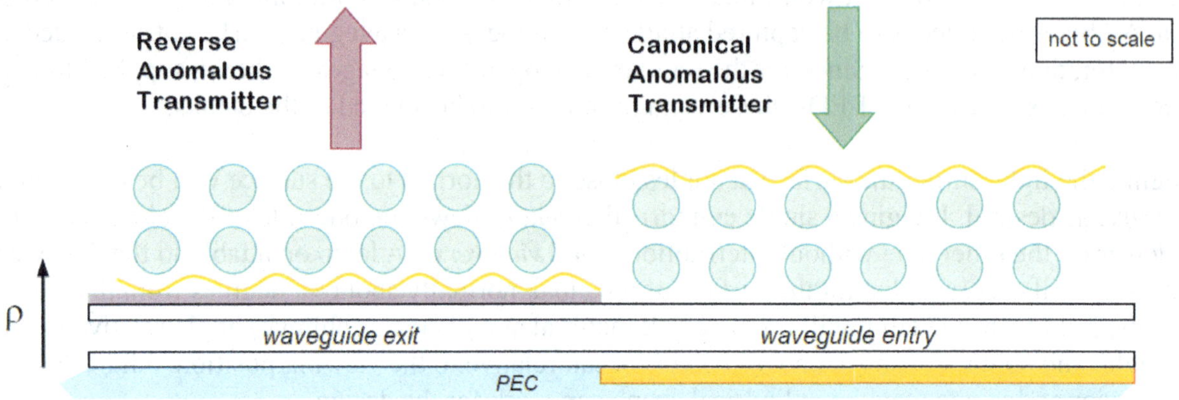

Figure 10. Top-down view showing very-thin cloak structure.

In this view the waveguide entry (alternately repeated around the cylinder circumference with the exits) is positioned below the Zhan transmitter. The gold backing is placed *behind* the waveguide. For the exits, the transmitter forms a common interface with the waveguide DPS interface (shown in purple, above). The outbound signal is picked up for offboarding by the bumpy gold surface that makes surface-plasmon contact with the DPS surface, leading to the dielectric outer layer. The outermost layer (not shown) is the glass of the display cube.

The problem of how to time the lightwaves to emerge "on the other side" is completely dependent on the materials used and their parameters. The design must ensure that the strongest signals persist up to a maximum distance of half a circumference before being released. A normal lens has a center of focus, but an NIM lens is *perfect* because all source positions are captured and faithfully reproduced at the exit points. This assumes the wave retains its original trajectory. The cloak, however, because it bends light, can't really retain the original trajectory. In other words, it seems unlikely that a light vector, after being captured by the waveguide, would have any memory of its entry angle when it exits. On the other hand, if the ports don't interfere with the homogeneous surface of the underlying waveguide, light *will* enter and exit. This implies that if the design works at all, the light will retain some kind of orientation.

Pitch dependence (PD) describes the incidence angles of light rays relative to the planes defined by two points on the cylinder's height axis and some radial point. Thinking of the internal waveguide surface as unbroken in both the radial and height directions, so that a ray entering at a high angle will spiral downwards, or a ray at a low angle upwards, then light rays at zero degrees will have the shortest duration of entrainment; while those at steeper angles will travel longer distances before emerging. Yaw dependence (YD) describes the incidence angles of light rays relative to a plane parallel to the cylinder's endcaps, at some height z. Light rays aligned with the angle of view will appear the strongest and have the longest duration of entrainment; while those at angles to the viewing angle, with the extreme being rays having entered ± 90 degrees to the viewing angle, will be weakest or most oblique and have the shortest duration of entrainment.

These angles naturally select the propagation direction. The angle of incidence affects the grating power, and so transmittance will have an associated power rating (very small), and this in turn will ensure that the light signal couples at the right time with the Zhan emitter. Weaker waves (as transformed by the media) will disembark earlier. The wave should exit the waveguide without losses from metal absorption, noise, and so on – but this is in an ideal world. Nonreciprocal media should help in this regard. A further challenge is the implicit preservation of the plane wave as it enters and exits. Plasmon coupling may degrade the signal, and the wavelength and angle at the incoming side may not be exactly duplicated in mirror fashion on the outgoing side.

In general the cloak has the following qualities:

- Dielectric spheres in metal (aluminum) in the outer wall are employed to couple waves into and out of the medium at all radial angles ([1], [2]);
- A bumpy metal surface (gold) at ingoing and outgoing boundaries is employed to catch radiation over a broad range of incident PD and YD angles ([2]);
- The bandwidth response is widened by stacking the spheres closely ([9]);
- Dielectric sizes are chosen to enhance surface plasmon modes;
- Negative index materials are used to propagate lightwaves in all directions ([20]);
- Effective medium theory determines the homogeneous parameters of the thin-layer design (including perhaps a gyrotropic nihility state [26]);
- Transverse-averaging handles layer-variation in the material parameters ([17]);
- Results are optimized for a given layer or assembly of micro-layers to find the best parameters; and indeed to find the best multilayer solution in general ([16], [26]).

The cloak is something that guides light. The nature of that guiding has different interpretations. The transverse electric field that travels around the cylindrical shell, as shown on the left of Figure 11 below, isn't really an accurate picture because the light is instantaneously moved through the metamaterial, as shown in the middle diagram for multiple light paths. The light path for a single incident angle is shown more accurately on the right. It turns out that all three depictions are valid.

Figure 11.

The display cube doesn't require a single cloak configuration with a single broadband response. Instead, since the cube is already split into color planes and spacer planes, and the cloaks are distributed to the spacer planes, the operating frequency of the cloaks populating a given spacer plane can be set to one of red, green, or blue. The operating frequency of the cloak is derived from the radial measures of the cylinder plus the surface nanogrid dimensions. Just as the color planes are fused by the eye into a single color, each associated spacer plane will be filled with cloaks tuned to the frequency of its adjacent color plane. The light bent by the cloaks, as a unit, will be blended by the eye into the final color.

15.10. Tuning

The element of magic that pervades the cloak design can be objectified with mathematics. That is, what words can't convey numbers certainly can. Getting lightwaves to enter and exit at the correct times and angles will come down to tuning the cloak parameters in just the right way. This is done with mathematical models.

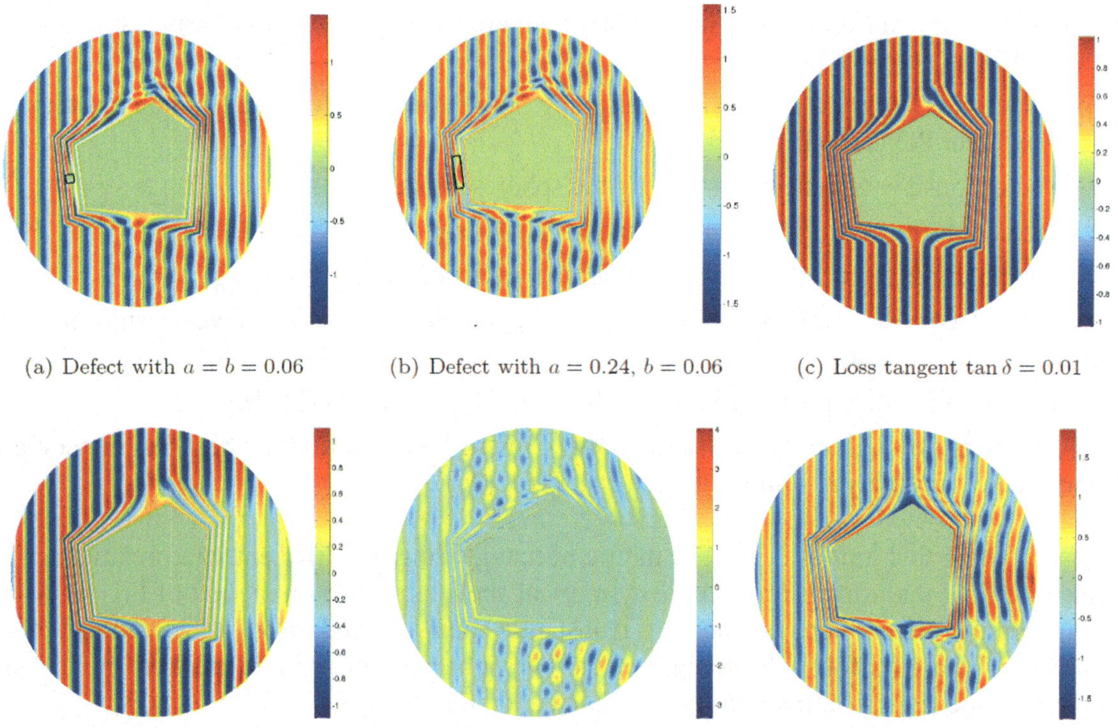

(a) Defect with $a = b = 0.06$ (b) Defect with $a = 0.24$, $b = 0.06$ (c) Loss tangent $\tan\delta = 0.01$

(d) Loss tangent $\tan\delta = 0.05$ (e) Dispersion with $k = 39$ (f) Dispersion with $k = 41$

Figure 12. (Courtesy [28])

This panel from a 2015 paper [28] shows various types of defects that can be simulated. This is useful because a cloak that shows defects in manufacturing can be simulated with mathematics to try to figure out what parameters or cloak locations are deficient. Items (a) and (b) show the effects of rectangular defects with sides of given lengths *a* and *b*. The electric field distribution is affected, more so for the larger defect, in both the forward scattering and back scattering directions.

Items (c) and (d) are simulated with a loss tangent that quantifies the absorptive property of the medium. The images show how, when the electric permittivity is changed, the medium becomes lossy and the forward scattering region is affected because the forward traveling wave power is absorbed by the lossy medium. The effect is barely perceptible in (c) but plainly visible in (d) on the back side. Since NIM slabs are lossy by definition, changing the loss tangent to a gain tangent could conceivably improve performance.

Finally, items (e) and (f) show what happens when the cloaking frequency, which for a perfect cloak must be fixed, is offset just a little from the proper solution. The perfect solution has a wavenumber k = 40, but

(e) shows k = 39, which is a disaster, and (f) shows k = 41, only slightly less dispersive. The vacancy on the back side in (e) is referred to as a forward scattering *shadow*.

The author of [28], the talented Li-Lian Wang, uses a mathematical model to simulate scattering through an *array* of spheres. This is extremely useful information to have because we need to know how the cloaks in a grid formation will affect each other. Normally this kind of computation is extremely expensive, but spectral element methods (SEM) are very fast and can handle this task. [29]

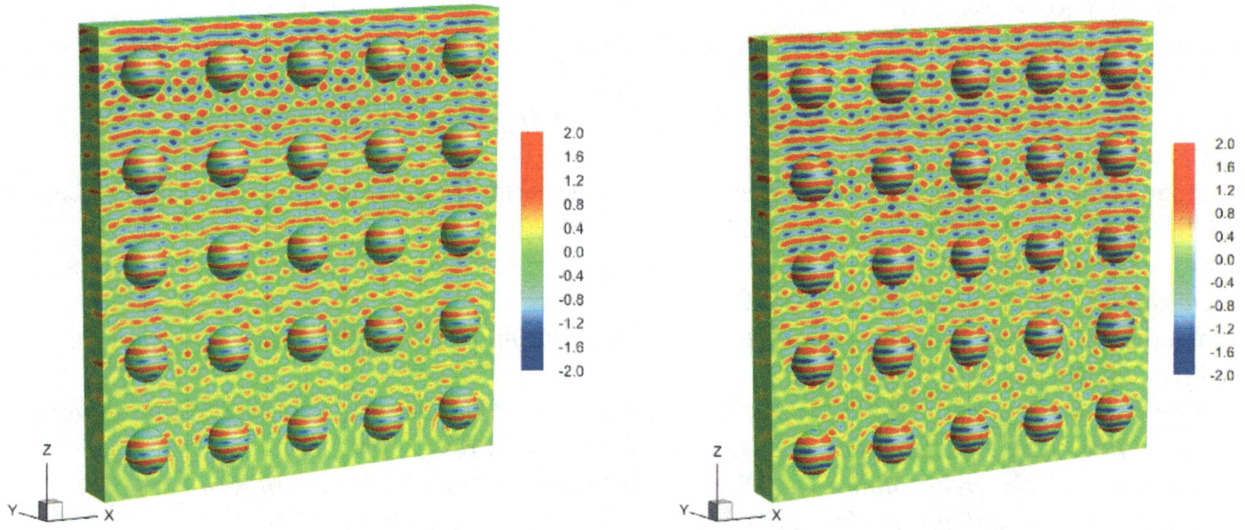

(a) Real part (with plane incident wave) (b) Imaginary part (with plane incident wave)

Figure 13. (Courtesy [29])

The images show the real and imaginary parts of a plane incident wave as it encounters an array of spherical obstacles. In the ideal world an array of cloaks will transmit light with no absorption and no reflection, with no scattering losses whatsoever, but in reality all of these things will play a part. The incident wave can be imagined as a lightwave, and the simulation shows how the light is attenuated as it travels from top to bottom. The top can be thought of as the cube perimeter, and the bottom as the cube interior. The behavior shown above can be referred to as the *dark forest effect* – that is, the cloaks lose effectiveness due to light dissipation due to scattering either by other cloaks or by semi-transparent electronics. The next chapter will present some possible solutions in this regard.

15.11. Conclusion

This chapter has explored invisibility technology with a focus on the freestanding cylinder, with its requirement of light conduction in addition to light transmission. A general design has been proposed.

How many years until two viewers can look at each other *through* an invisibility cloak? Carpet cloaks already work but at very small scales. In 2015 it was predicted by UC Berkeley researchers that large-scale carpet cloaks could be achieved by 2020. That turned out to be optimistic. A cylindrical cloak is much harder because light isn't just reflected, it's bent. 2030 might be a realistic estimate for a fully-working cylindrical solution in the lab.

References

[1] Liu, Zhan, Chen et al. *Dual broadband near-infrared perfect absorber based on a hybrid plasmonic-photonic microstructure*, Optical Society of America, 2013
[2] Zhan, Wang, Dong et al. *The Anomalous Infrared Transmission of Gold Films on Two-Dimensional Colloidal Crystals, Advanced Materials*, v. 18, pp 1612 – 1616, 2006
[3] Steshenko et al. *Super-Resolution and Near-Field Enhancement with Layers of Resonant Arrays of Nanoparticles*, in *Applications of Metamaterials*, Capolino (Ed.), CRC Press, 2009
[4] Sihvola, Ari. *Polarizability of Simple-Shaped Particles*, in *Theory and Phenomena of Metamaterials*, Capolino (Ed.), CRC Press, 2009
[5] Alu and Engheta. *Scattering Cancellation and Plasmonic Cloaking*, in *Applications of Metamaterials*, Capolino (Ed.), CRC Press, 2009
[6] Choy, Tuck C. *Effective Medium Theory: Principles and Applications*, Second Edition, Oxford Science Publications, 2016
[7] Yaghjian et al. *Electromagnetic Cloaks and Concentrators*, in *Applications of Metamaterials*, Capolino (Ed.), CRC Press, 2009
[8] Martin-Moreno and Garcia-Vidal. *Enhanced Transmission at Optical Frequencies*, in *Applications of Metamaterials*, Capolino (Ed.), CRC Press, 2009
[9] Garcia de Abajo, Gomez-Santos, et al. *Tunneling Mechanism of Light Transmission through Metallic Films*, Physical Review Letters, 95, 2005
[10] Steshenko and Capolino. *Single Dipole Approximation for Modeling Collections of Nanoscatterers*, in *Theory and Phenomena of Metamaterials*, Capolino (Ed.), CRC Press, 2009
[11] Steshenko et al. *Super-Resolution and Near-Field Enhancement with Layers of Resonant Arrays of Nanoparticles*, in *Applications of Metamaterials*, Capolino (Ed.), CRC Press, 2009
[12] Simovski and Tretyakov. *Material Parameters and Field Energy in Reciprocal Composite Media*, in *Theory and Phenomena of Metamaterials*, Capolino (Ed.), CRC Press, 2009
[13] Vallecchi and Capolino. *Metamaterials Based on Pairs of Tightly Coupled Scatterers*, in *Theory and Phenomena of Metamaterials*, Capolino (Ed.), CRC Press, 2009
[14] Iyer and Eleftheriades. *Fundamentals of Transmission-Line Metamaterials*, in *Theory and Phenomena of Metamaterials*, Capolino (Ed.), CRC Press, 2009
[15] Qiu et al. *A Review of Chiral and Bianisotropic Composite Materials...*, in *Theory and Phenomena of Metamaterials*, Capolino (Ed.), CRC Press, 2009
[16] Zhang and Wu. *Cylindrical Cloaking at Oblique Incidence With Optimized Finite Multilayer Parameters*, Optical Society of America, 2010 (online)
[17] Silveirinha and Fernandes. *Transverse-average field approach for the characterization of thin metamaterial slabs*, Physical Review E 75, 2007
[18] Silveirinha, Mario. *A Metamaterial Homogenization Approach with Application to the Characterization of Microstructured Composites with Negative Parameters*, Physical Review B, 2007 (preprint online)
[19] Baccarelli et al. *Modal Properties of Layered Metamaterials*, in *Theory and Phenomena of Metamaterials*, Capolino (Ed.), CRC Press, 2009
[20] Yao et al. *Three-dimensional nanometer-scale optical cavities of indefinite medium*, PNAS v. 108, 2011
[21] Marques and Martin. *Split Ring Resonators and Related Topologies*, in *Theory and Phenomena of Metamaterials*, Capolino (Ed.), CRC Press, 2009
[22] Alu and Engheta. *Negative Refraction in Infrared and Visible Domains*, in *Theory and Phenomena of Metamaterials*, Capolino (Ed.), CRC Press, 2009
[23] Moura, Tiago. *Metamaterials and Double Negative (DNG) Media*, undated and unsourced, probably 2009

[24] Burghignoli et al. *Enhancement of Directivity by Using Metamaterial Substrates*, in *Applications of Metamaterials*, Capolino (Ed.), CRC Press, 2009
[25] Liu, Kexin. *Theoretical Investigation on Propagation and Coupling of Nonreciprocal Electromagnetic Surface Waves*, Doctoral Thesis, KTH Royal Institute of Technology, 2016
[26] Tuz, Vladimir. *Polaritons dispersion in a composite ferrite-semiconductor structure near gyrotropic-nihility state*, Journal of Magnetism and Magnetic Materials, 2016
[27] Lee, Hyesog et al. *Super-Resolution Imaging with Hyperlens*, in Applications of Metamaterials, Capolino (Ed.), CRC Press, 2009
[28] Wang, Li-Lian et al. *Seamless Integration of Global Dirichlet-to-Neumann Boundary Condition And Spectral Elements for Transformation Electromagnetics*, 2015
[29] Wang, Li-Lian et al. *Accurate Calculation of Spherical and Vector Spherical Harmonic Expansions via Spectral Element Grids*, 2017

Chapter 16
Viewing Hemisphere

16.0. Introduction
In this chapter the subject of the viewing hemisphere will be discussed. The viewing hemisphere envelopes and surrounds the core cubic display, allowing an uninterrupted three-dimensional viewing experience without the distracting shifts in perspective associated with a clear rectangular solid.

The viewing hemisphere necessarily complements the invisibility planes of the 3D display device. Though metamaterial cloak design has become a science in itself over the last twenty years, the best designs are calculated with plane-polarized lightwaves only. This isn't a fatal limitation because light can be polarized before it reaches the core cube – and this will be done by the viewing hemisphere.

16.1. Discussion
The scale of the nano-structures in the display cube requires some innovation in etching or printing the parts. The discussion in this case parallels the discussion of "how small is too small" for regular microchips. Is there a lower limit beyond which etching technology cannot go? In the 1980s experimental circuits were etched with such tiny lines that quantum fluctuations began to interfere with their reliability.

By using X-ray etching techniques it's possible to cut channels 0.1 microns wide – only 1000 times bigger than atomic size. A small particle of radioactivity at this scale might flip a tiny transistor from the "off" to the "on" configuration. An invisibility cloak, whether active or passive, doesn't actually perform any calculations like a computer would, but interference can take other forms.

The important point here is that X-rays are used in the technology of extreme miniaturization for a reason: at nano-scales the photo-etching techniques used to engrave dies and wafers can't use visible light. With wavelengths of half a micron, visible light is too thick! It would be like Leonardo da Vinci trying to paint the Mona Lisa with a house roller. X-rays, with their smaller wavelengths, are used instead. An even smaller technology is called electron beam lithography, but it has stability problems. X-ray lithography is more stable but the etching is very slow – sometimes taking hours. Speed and stability are two action items that need to be addressed for this etching technology to become a production reality.

Some very small pieces etched with deep X-ray lithography. [1]

Invisibility cloaks depend on two design factors: a) the layout of the construction; and b) the metamaterial properties which are tailored for that construction. The materials have to be manufactured with exact property values that match the mathematical results of the invisibility calculations.

The entire 3D display device is enclosed in glass but the glass enclosure won't be invisible: the glass will appear as glass when held in the hand. Having an outer shell of glass should simplify the cloak design since the refractive index is linked with physical density. To be invisible in air the cloak material would have to be as light as air to avoid refraction. This isn't a practical possibility. But glass is much denser than air and so opens up more design choices for various metamaterial configurations.

The basic cloak calculation is based on a Maxwellian light-conducting transformation that determines the cloak's material properties. In other words, the mathematics dictate the form. In terms of refraction, the closer the actual materials come to matching their boundary conditions, the simpler the configuration can be.

Assuming a higher density due to glass boundaries instead of air, there are already designs that achieve a very low *scattering efficiency* – this last being an indicator of how much light is scattered at the cloak surface. A lower number is better, and a nearly zero scattering efficiency can be achieved with a cloak design of four concentric cylindrical layers defined by six parameters. [2] This design works for plane-polarized lightwave incidence-angles between +/– 50 degrees.

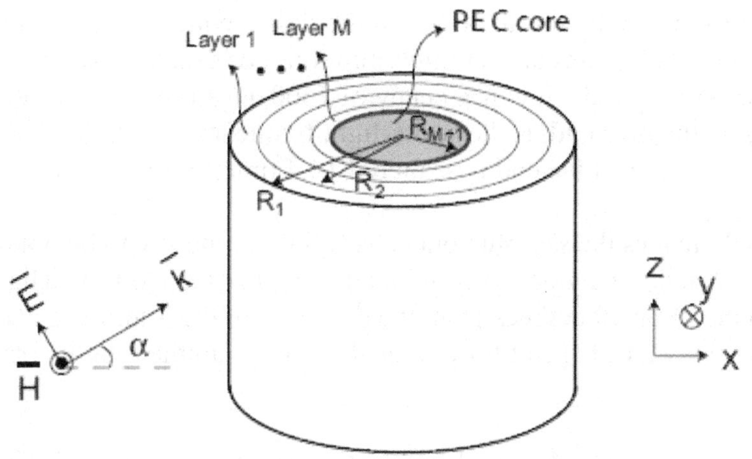

The four-layer design. (Courtesy [2])

16.2. Polarization Considerations

The properties of ordinary glass are revealed by a simple experiment. A beam of light which has passed through a polarizing disc and is vibrating north and south, will be unable to pass through a second disc which is set with a vibration direction at right angles to this, i.e. east and west. This is a system of "crossed polarizers". If a piece of glass is inserted between two crossed polarizers, the darkness of the field will remain unaltered since glass, being amorphous, imposes no change in the vibration direction of light from the first disc.

Now, put a crystal lattice between the polarizers and things are less straightforward. A *uniaxial* crystal like quartz has double refraction which corresponds to an *extraordinary ray* which moves, when the crystal is rotated, toward and away from the immovable *ordinary ray*. Other minerals with other lattice systems have *two* extraordinary rays. Optically, a grid seen through the quartz will become doubled as the quartz is rotated, and then go back to a single image with further rotation around the axis.

When quartz is rotated between crossed polarizers, the net result as seen from one side or the other of the line of sight through the polarizers is four positions of darkness or *extinction* at 90 degree intervals, and four positions of maximum brightness, offset by 45 degrees from the extinction rays as the quartz is rotated.

It's interesting to note that glass under strain, which happens when the glass has been badly annealed, shows "anomalous" double-refraction between crossed polarizers: there's no regularity in the pattern because the maxima and minima are random.

The absorption of light at certain angles is the basic idea behind polarization, and the ability of a lattice structure, as in a crystal, to bring this about is the inspiration behind old-style polarizers like polaroid coatings, which use iodine crystals; and more modern productions like Corning Polarcor. Truly amorphous glass is no hindrance to a lightwave that's been polarized. The wave will retain its polarization despite the interference of borosilicate atoms.

16.3. Cloak Considerations

Ambient light is ambient because it arrives from all directions. Rayleigh hypothesized that the scattered power of sunlight is proportional to the fourth power of the light frequency. When light passes through a gas of neutral atoms, the bound electrons are set into vibration. For example, suppose white sunlight has a high frequency of 3900 angstroms and a low frequency of 7600 angstroms. The ratio of the two, with the resonant frequency being on the blue end, is the white-light frequency or (7600 / 3900), and the fourth power of this fraction is 14.4. So the blue end is scattered 14.4 times more than the red.

Rayleigh scattering not only makes the sky blue but causes light in the sky to be polarized. The polarization of the light of the blue sky can be easily observed with a piece of polarized film. The maximum amount of polarization is found in a direction 90 degrees from the direction of the Sun. Bees use this fact to navigate from hive to flowers on a sunny day. [3] But for practical purposes ambient light arrives from every direction.

When light hits the cloak at an angle normal to the cylinder axis the horizontal polarization is *decoupled* from the vertical polarization. Light travels in orthogonal waves, and in the Maxwellian world one plane represents the electric field, and the other the magnetic field. A decoupled design is convenient but represents only the zero-degree normal and so accounts for a very small number of the arriving waves. Coupling of the vertical and horizontal polarized fields takes place when the light arrives at a non-normal angle to the cylinder axis. A design that handles coupling requires more degrees of freedom in the material parameters.

There are three ways of describing light:

1. Because of multiple collisions over time there's no single state of polarization for ambient light. This is the "outermost" state of light before it reaches the cloak.
2. Light can be polarized with a fixed orientation so that, for example, electric or magnetic fields are polarized exactly horizontal or vertical relative to the cloaking columns.
3. Usually one speaks of one direction of polarization, say the vertical. But light will arrive at different vertical azimuths, at angles above and below the horizontal.

All three of these considerations are important for the design of the viewing hemisphere. Obviously the light-bending effectiveness of a cylindrical cloak will degrade as the light angle becomes more acute. There are two reasons for this. First, the material construction is limited by its mathematical definition; and second, the energy distribution on an intercepting surface depends on the angle of the incoming energy rays. The energy distribution has the highest concentration per unit area when the arriving rays are perpendicular to the surface.

One might add a fourth item, being the viewing angle of the cloaks, being completely independent of the light angles. As the viewing angle increases the columns will tend to vanish anyway because you're looking at them end-on. Ideally, as the columns are rotated into their full extension, they will be fully invisible along with the internal circuitry they hide.

The key consideration is that invisibility degrades at acute *light* angles, and to account for this a coupling design becomes desirable.

Another consideration behind many cloak designs is the notion of a *perfect electric conductor* (PEC) core. This means that any external current is essentially reduced to zero at the core. The metamaterial construction cancels out any incident magnetic and electric field oscillations. For light radiation the strength of these fields will be extremely weak. Whether a PEC core can be achieved or not depends on the quality of the construction and on the performance of the component materials. The cloak's efficiency will determine the degree of destructive cancellation of the majority of incident radiation.

The six parameters for each of the four cylindrical layers are values for electric permittivity (ε) and magnetic permeability (μ). The material defined by these six values is *homogeneous* for each layer. (Homogeneity is explained in Section 7.) The adjacent layers produce a scattering efficiency of close to zero. How does this work? Normally, plane-polarized radiation scatters electrons in the material. Photons are not electrons, but photons can displace electrons since light particles have momentum. Though the magnitude of the scattering is tiny, it's nonetheless a widespread phenomenon in nature. The magnetic field is assumed to be polarized along the z (vertical) axis, and the electric field in the x,y plane. As mentioned, the electric field is very weak, which is desirable because when a cloak is activated with an applied electric field, something else happens. The electric field entrains the light. This entrainment is called the *Kerr electro-optic effect* and happens because the molecules in the medium (such as a crystal) are aligned due to the electric field. The medium behaves like a uniaxial crystal in which the electric field defines the optic axis. The four layers described in [2] achieve a zero current at the core due to destructive interference of the complementary waves associated with vertical polarization, and take into account a non-normal incidence of +/– 50 degrees for the arriving wavefronts. This behavior is purely theoretical but the simulation is convincing.

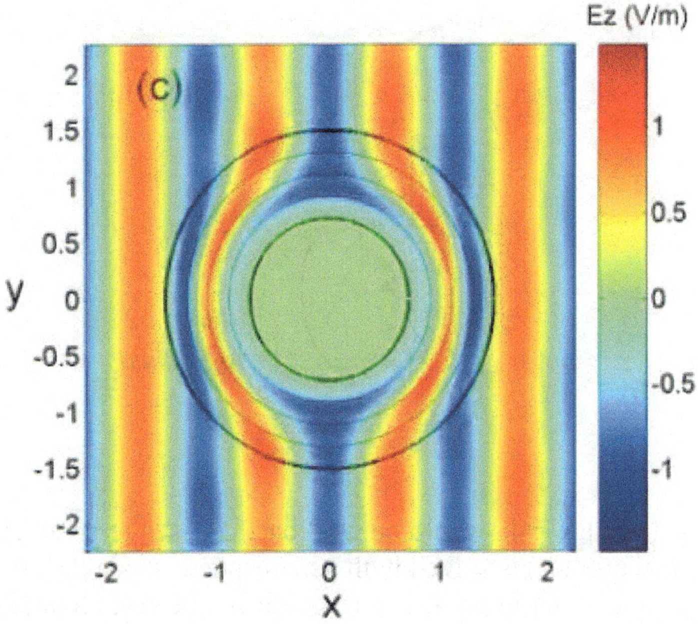

The optimized cloak with a scattering efficiency of 0.013. [2]

16.4. Corning Polarcor

The viewing hemisphere is the first line of defense against unpolarized ambient light. The main idea is that the arriving wavetrain can be polarized by the material it enters. Granted, some arriving light will be lost at the glass boundary, but once inside the glass, light is polarized and will travel through the glass until it reaches the cloaks.

Gemcutters know that the cut of an aquamarine is best when the table (the top surface of the gem) is perpendicular to the growth axis of the natural beryl. [4] This is because aquamarine is *dichroic*. The crystal is doubly refractive, and as light enters the medium it's slowed down. But it's slowed down along two different axes, so it takes two different amounts of time to be reflected out of the lattice. When you turn the crystal one way or the other, more blue or more white light is apparent. Dichroism can be described as a type of colored polarization.

Corning Polarcor [5] is a unique material that mimics this behavior, except that it's singly refractive. Corning insists on calling their material a dichroic polarizer, but really it's monochroic. In other words, the material selectively absorbs light which vibrates in a particular plane relative to the material's crystalline axis. Polarcor is anisotropic for this reason. More, it's an "absorption polarizer."

Polarcor is a borosilicate glass that contains titanium, aluminum, and silver. The polarizing mechanism consists of elongated silver crystals in the outer 20 μm to 50 μm of the glass surface. The lattice of silver produces polarization in a preferred direction. The material can be cut so that the angle of polarization is whatever the customer wants.

Polarcor is a remarkable specialty material, but it's only a start. It has some limitations. First, the material is meant to be sandwiched between, for example, the plane ends of two fiber-optic cables, which means that the angle-of-incidence is limited to 0 degrees (the normal) +/– 5 degrees. For a fiber-optic source, the arriving light direction is fixed; but ambient light arrives from many different directions. The Polarcore slice would, if light was arriving from the left at a high incident angle, exhibit the behavior sketched below.

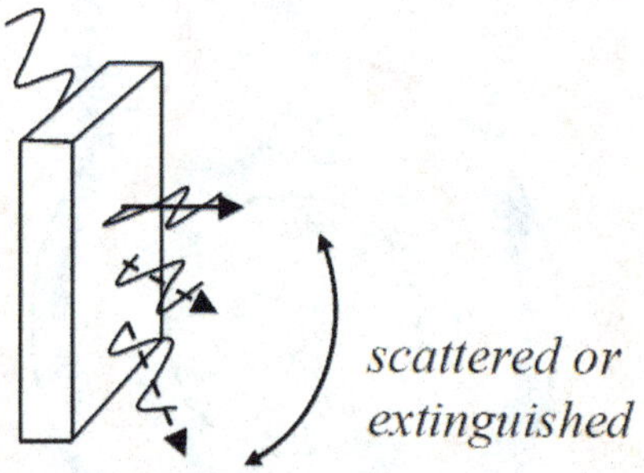

With increasing incidence angles the light would be weakened and scattered after passing through the material. This incidence-angle problem is solved with a hemisphere, since the silver particles can be aligned with the hemispherical contour to produce strong transmission at all ambient angles. This is sketched below.

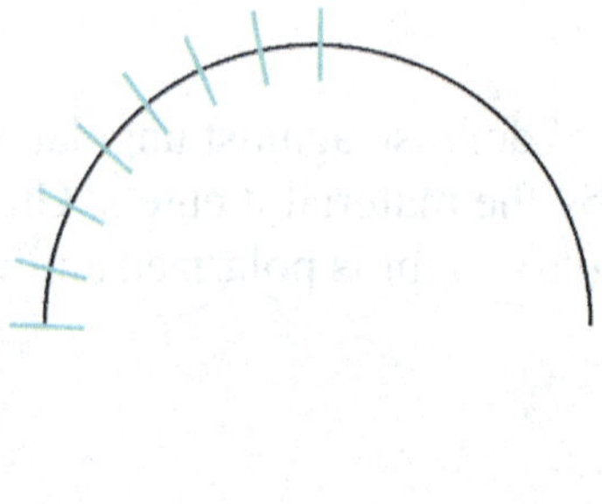

Think of this as a two-dimensional cross-section of the hemisphere. Light from every angle will be plane-polarized at that angle, and every angle of the entire hemisphere will be polarized along the vertical. The upper reaches will have very little impact on the vertical cloak columns because the columns themselves are restrictive above +/– 50 degrees, but the light transmission will still be maximal for those high incidence angles and thus provide ambient (though polarized) illumination for the voxel elements.

There are other problems with Polarcor – ones that aren't so easy to work around. The principal limitation of the material is its operating wavelength. Visible light ranges from 380 nm (nanometers, violet) to 800 nm (far red), but the Polarcor product line only starts at 633 nm. There's a very good reason for this: the silver particles are harder to manufacture at higher frequencies. The display cube requires the full spectrum, so this limitation will have to be overcome in the future. Unfortunately, there's another problem: at visible-light frequencies the Polarcor material becomes *photochromic*, which means it darkens. This is because visible light heats the silver particles and causes them to expand, thereby changing the lattice configuration. This problem will also need to be overcome.

Corning flatly states that Polarcor will not be made for the full visible spectrum. Transmittance would drop below 75 percent. Corning offers a wideband variation that spans 600 nm to 1100 nm (from visible to infrared) and its transmittance (with no AR) is 60 percent.* The frequency bandwidth is 500 nm, but this is a nominal range since near-zero reflectance occurs only between 1075 and 650 nm, giving an operational range of 425 nm. By comparison, the range of visible light is 800 – 380 = 420 nm: about the same bandwidth but shifted towards the higher frequencies. If manufacturing can be improved, a wideband Polarcor-type material could be made for the visible spectrum. It turns out that higher frequencies require a thicker piece of glass – probably because additional layers of interlaced silver particles are required to absorb the higher frequencies. Polarcor's standard thickness is about half a millimeter; lower-bandwidth UltraThin material has a thickness of 0.03 mm.

Transmittance is related to *insertion loss*. When insertion loss increases, transmittance decreases. Insertion loss increases when thickness increases. So a thicker lattice for higher frequencies will also reduce transmittance. The table below gives some insertion loss values for two extreme wavelengths for the standard thickness of 0.5 mm.

coating	633 nm	2100 nm
AR	0.76	< 0.06
No AR	1.16	< 0.43

Table 1: Insertion Loss

Transmittance (affected by insertion loss) for these two cases is 76.5 percent (633 nm) and 90.5 percent (2100 nm). So clearly the infrared is easier to transmit than the visible.

The physical layout of Polarcor is a 50 μm active polarizing layer, a 400 μm glass core, and a second 50

* An anti-reflective (AR) coating improves light transmittance because there's less reflection at the glass surface. Any improvement is desirable because there's no way with efficient polarizers to avoid lower transmittance at visible wavelengths. The more light coming in the better.

μm active polarizing layer (at the same polarizing angle). The entire thing is a solid piece of borosilicate glass. Two sides are used purely to increase the polarizer efficiency, and thus what is known as the *contrast ratio*. A contrast ratio of 10,000 to 1 is extremely good and when light is passed through crossed polarizers the light should be nearly extinguished. When the ratio goes down, the near-black contrast between crossed polarizers turns grayer because the polarization efficiency is worse and unpolarized light leaks through.

Polarcor has a *Brewster angle* of 56.3 degrees. The Brewster angle is directly related to polarization and to a material's *refractive index*. For an air-glass interface, the Brewster angle is in fact 56 degrees. This angle has a few special properties. Define the *plane of incidence* as the plane formed by the incoming ray and the normal to the glass surface.

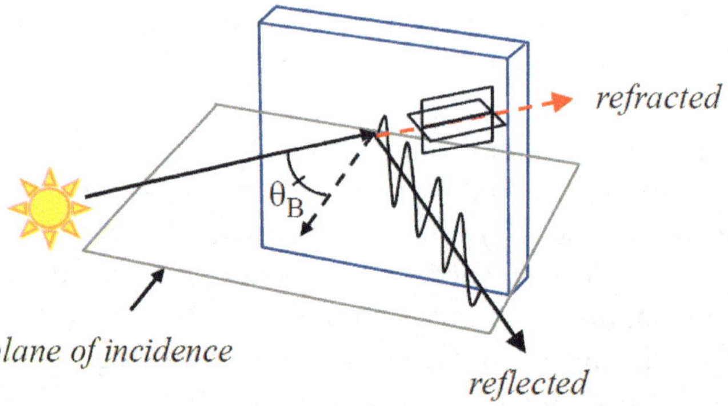

A plane harmonic wave arriving at the Brewster angle (56 degrees) relative to the surface normal (0 degrees) will be reflected with a linear polarization that's perpendicular to the plane of incidence. However, the amount of reflected light is small. The ray is also refracted (into the glass) at ninety degrees to the reflected ray, with radiation both parallel and perpendicular to the incidence plane.

For a plane wave at 56 degrees from the normal, the reflected light is a weak polarized light and so the reflectance isn't zero. Some devices produce a strong polarized light by stacking glass plates to *amplify* the reflected polarization – much more light is transmitted from layer to layer than reflected, but the reflected light adds up. Three plates will produce three times the reflected polarized light.

Zero reflectance can only be achieved if the incoming ray is *already* polarized parallel to the incidence plane. A *Brewster window* can be set up whereby a beam of light that's linearly polarized in the transverse magnetic mode is incident at the Brewster angle on a glass plate with parallel faces. In that case there's no perpendicular component to be reflected. No light is reflected from the first face and there is no internal reflection from the second face. The light is entirely transmitted. It's a *perfect window*.

Angles between the normal (0 degrees) and 56 degrees all have some reflectance; the reflection coefficient increases dramatically from 56 to 90 degrees, that is, when the incident light arrives at a glancing angle to the surface. As well, the linear polarization of the reflected ray decreases on either side of the Brewster angle. Since Polarcor is glass with a refractive index of 1.52, and air has a refractive index of 1, the Brewster angle is 56 degrees. Other Brewster angles are possible. The Brewster angle depends entirely on the refractive indices of the contact media: it's just an inverse tangent of the refractive index ratio of the two optical media. Glass against glass gives a Brewster angle of 45 degrees because the refractive index ratio is unity, and the arctangent of 1 is 45.

16.5. Viewing Hemisphere Challenges

Here are some challenges related to the viewing hemisphere.

<u>1. Overcoming the Dark Forest Effect</u>

Ideally, the invisibility cloaks should bend 100 percent of the arriving light and so act as "perfect transmitters." In reality, some arriving light will be attenuated, leading to a shadowy area near the center of the display volume. The viewing hemisphere may help since light will reach the interior through the top of the dome, and though the incident angle is high, light in the central region will help illuminate the LEDs. The light arriving from all sides will be vertically polarized so that the outer parts of the display volume will achieve full invisibility leading to an open viewing area in the core region. The hope is that interpenetrating light rays at spherical angles will combine to render the Dark Forest Effect negligible. (Read on.)

<u>2. Combatting weak color saturation in ambient light</u>

To cut down on power requirements, which otherwise would drain the batteries when the cube is undocked, ambient light compensates for LED power. Remember there is no "backlight" in a 3D display. But ambient light alone may not be strong enough to prevent LEDs from looking washed out. How can the saturation be boosted while keeping battery usage to a minimum? One solution complexifies the design but may be worth it. If the core region is defined as the inner third of the volume in the horizontal directions, and two-thirds in the vertical, then more power can be applied to the core region. The LED brightness would be increased electrically. This would help illuminate the interior LEDs and wouldn't affect the cloaked columns.

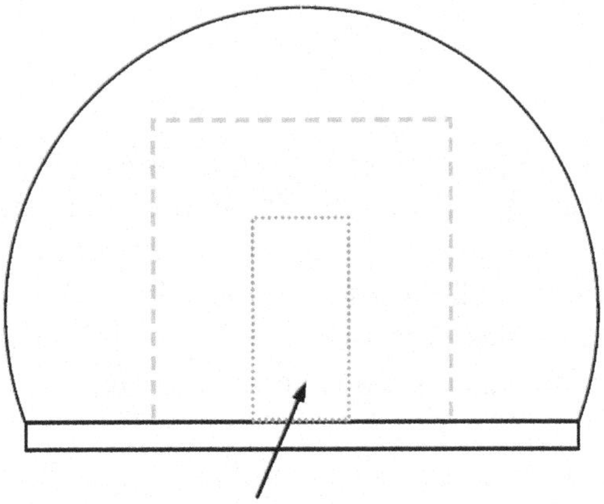

possible attenuated region

Another solution involves the Brewster angle. Light admitted from the top of the dome will penetrate to the bottom of the cube because the cloaks are arranged end-on relative to those higher angles. Since the 3D array is regular, there will also be open corridors where light from opposite sides of the hemisphere will penetrate. Whether from the top or from opposing sides, the idea is to reflect the light, with polarization, at an angle so that it reaches the invisibility columns. Putting a silvered mirror on the zero-level of the array would be distracting and would reflect the columns themselves. A better solution is micron-thick glass layers stacked at an angle, that, while not reflecting all of the light up into the array, would reflect a good percentage – with polarization according to the Brewster angle, and without being visually distracting when looking at the reflectors with the naked eye. The final reflector design will require some thought. One solution would be similar to a concave circular reflection grating. It would rest on the base plane and reflect light upward.

3. Unwanted iridescence at visible-light wavelengths
The fit and finish of the cube's interlocking pieces, some of them at the micron scale, must be flawless or else, like the structures in a peacock's feather which cause iridescent effects due to refraction, the display cube will become iridescent. The structures in feathers and insect shells are so small they can only be seen with an electron microscope at about 10,000 times magnification. Iridescence is also produced by thin films, as in soap bubbles and oil slicks. [3]

4. Computer-controlled glass fusing of die layers
The process of making 3D laser-etched images in a glass block, which is the basic idea behind fusing the cube, needs attention. First, the laser-based fusion of die layers requires a laser beam one hundred times more accurate than that used for 3D glass artwork. Second, it turns out that the laser-etch causes the glass to become brittle. The display cube must not fracture and should last for years under constant use.

5. Reducing lensing distortion of the display image through the glass
The idea of embedding the cubic display volume within a glass viewing hemisphere has yet another problem: the lens-like distortion due to the sphericity of the hemisphere itself. Though infinitely better than the distortion of a cube of glass, or even a rounded cube of glass, the edges of the hemisphere act as a lens. A mathematical counter-distortion of the image in real time would solve the problem, provided that: a) a built-in retina locator could pinpoint the viewer's eye position; and b) the cube's yaw and tilt relative to some baseline could be quickly determined via sensor software. Until the bandwidth for this kind of processing is available, the consumer will have to live with some distortion of the image near the periphery of the globe.

16.6. Parts Explosion

Even though there are no moving parts in the display cube, it nonetheless contains many parts. Most of the parts will be *abstract elements*, meaning that they'll exist in digital memory and take the form of graphical elements that ultimately get etched into a die or wafer. The number of actual separate physical parts is far exceeded by the abstract elements of which they are composed. For a low resolution of 128 voxels cubed, the number of parts including abstract elements, but counting each invisibility cloak as a single part, is about 56 million. This isn't as fearsome as it sounds because the design is repeated from layer to layer.

Each column is composed of 384 interconnected cloaks. And each individual cloak, whether the design of the previous chapter or the four-layer approach in this chapter, will be prepared with laser deposition and electron-beam lithography. The fit and finish of these interlocking pieces must be superb. The final assembly of the whole cube, once everything is etched and printed, will consist of more than 5 million separate physical parts. Packing them into a durable consumer device that fits in the palm of your hand will be a mighty challenge indeed.

16.7. Effective Medium Theory

To see how extensive the Dark Forest Effect really is it might be possible to mathematically simulate the ambient illumination of the cube array entirely in software. (See the SEM examples in Chapter 15.) One well established modeling technique for multilayer cloaks is Effective Medium Theory (EMT). To better understand the four-layer cloak we need to define some terms.

First, homogeneity can mean different things in different contexts. For example, optical homogeneity describes a medium like glass that has a uniform refraction coefficient throughout the body of the medium. The glass can lose homogeneity if the annealing process is faulty. At the strictly atomic level, homogeneity of a material means that the number of electrons per unit volume is constant throughout the material. But for effective medium modeling, homogeneous behavior occurs when the structural length scale of the material becomes much smaller than the wavelength of the electromagnetic radiation propagating through it. Under this condition the metamaterial will behave like a homogeneous medium characterized by just two (in the absence of magnetoelectric coupling) frequency-dependent parameters: the effective permeability (μ), and the effective permittivity (ε). Notice these are different than the base permeability and base permittivity discussed previously. (See Section 3.)

Both effective parameters may be tensors if the material is anisotropic. One of the key benefits of metamaterials is that there's full control over both the magnetic (permeable) and electric (permitted) response (in principle), so that achieving a metamaterial with a large wavelength to unit cell size ratio becomes possible. The current state of the art is a ratio between 70 and 200 for microwave metamaterials in the gigahertz range. [7]

Periodicity in the design of metamaterials is a convenience but not a necessity, since metamaterials in general exhibit a macroscopic response that depends exclusively on the nature of their molecules in the frequency range of interest. The utility of periodicity outweighs its complexity, since many applications, including invisibility, are periodic. No matter what kind of mathematical model is dreamed up, periodicity must be built into the final manufactured product. Therefore periodicity, which can manifest in one, two, or three dimensions, is convenient for two reasons: 1) Fabrication-wise, structures composed of identical cells are easier to engineer; moreover, higher metamaterial homogeneity can be achieved if all cells are identical, because the upper bound acceptable for the distance between molecules is then fixed – it doesn't change since the lattice is regular; 2) Computation-wise, periodic structures are most conveniently analyzed using traditional periodic structure theory; moreover, this theory describes accurately the behavior of the metamaterial structure not only within the metamaterial wavelength range, but outside of it, which is useful in practice to account for anisotropy effects. [8]

Effective medium theory pertains to analytical or theoretical modeling that describes the macroscopic properties of composite materials. At the constituent level, the values of the materials vary and are inhomogeneous. Precise calculation of the many constituent values is nearly impossible. However, approximations can be produced which describe the properties of the composite material as a whole. A composite can have multiple dielectric layers. Each layer is considered to be homogeneous and isotropic. The following explanation applies to flat layers, but the idea can be extended to very small surface areas on a cylinder.

Homogeneity has a precise meaning when designing a composite material, in that the same scattering parameters S can be defined according to effective medium theory for reflection and transmission. Individual permittivity and permeability values for each layer can be modeled as a single *effective* permittivity and a single *effective* permeability. In other words, the multi-layered structure can be abstracted as a one-layer homogeneous medium with permittivity and permeability. The scattering problem for the original multi-layered substrate can be solved with a method of integral equations to get the S parameters.

Teng Zhao at the University of Iowa used custom software to do this, an analyzer called Equivalience Principle And Connection Scheme (EPACS) which avoids computation of a multi-layer periodic Green's function, and so is very efficient. Once the S parameters are known, the effective permittivity and permeability can be calculated. The effective composite approach is shown below. Variable *d* is the total thickness of the structure.

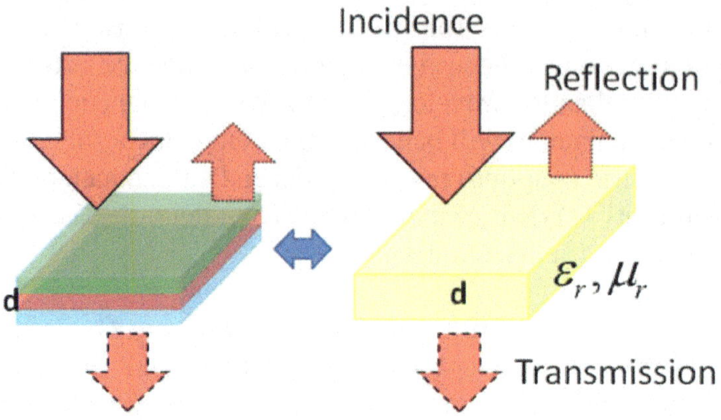

(Courtesy [9])

One effective medium modeling approach has a constant propagation, and a second approach a constant reflection coefficient. These two models give different results for effective parameters. For the general case, it's more accurate to use the tensor form for effective permittivity and permeability in the effective medium model. For each layer, as shown below, the material is homogeneous and isotropic. However, the effective medium is anisotropic. Permittivity, permeability, and thickness is known for each layer *i*. When one homogeneous effective layer is used to replace the multi-layer structure, the tensor forms of the effective parameters (right-hand side of diagram) give better results because they can model different wave behavior.

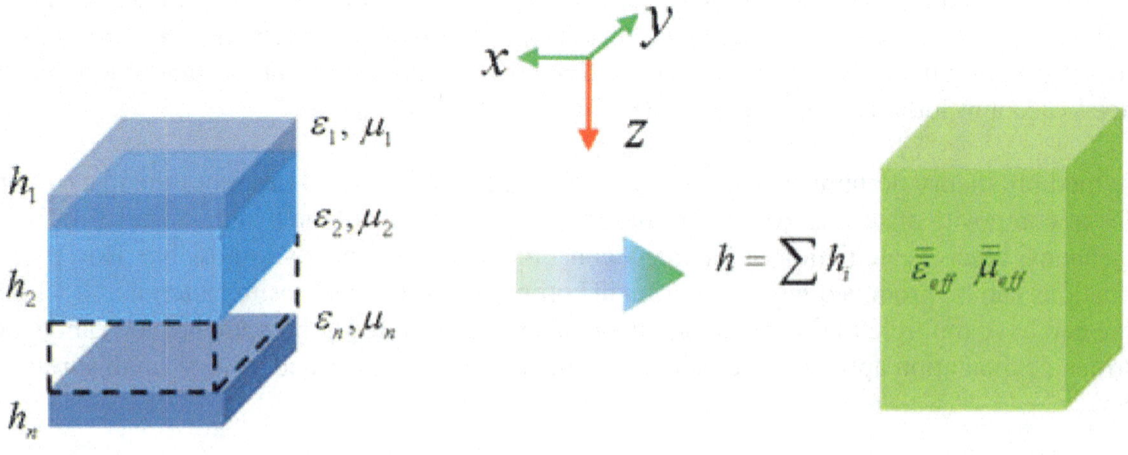

(Courtesy [9])

Tensors are particularly convenient when the effective permittivity is *uniaxial anisotropic*. Tensors provide a symmetry which is missing when one parameter or another is held constant. For example, when the electric field E is *perpendicular* to the interface, the results simplify to a simple effective permittivity equation at a low frequency. For an electric field E *parallel* to the interface, the simplified equation is nearly identical to the perpendicular result except for a single inverse term. Presumably the symmetry that comes with tensors applies as well to effective permeability at the two electric-field angles. The model is uncoupled: less well behaved electric-field angles with respect to the dielectric layers would introduce coupling and thus the model would be more complicated. The diagram below shows two different effective medium models, the only difference being the wave propagation direction *k*. Both models use the same scattering coefficients S.

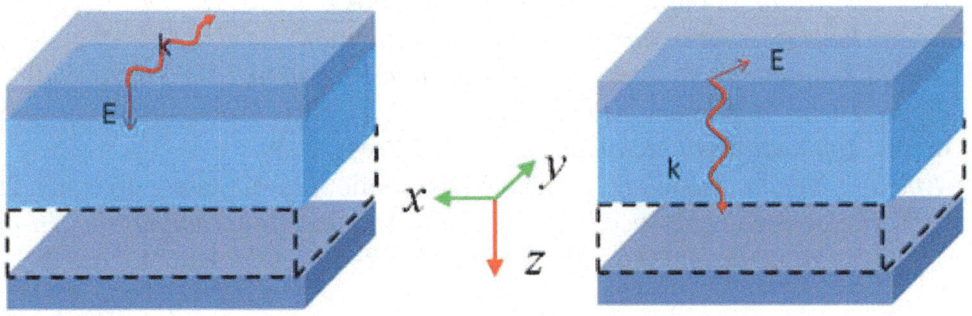

(Courtesy [9])

Note that effective medium theory is purely a computing convenience. If someone wanted to dispense with EMT, a model could be made for the individual layers and computed that way, but the model in that case would be more complicated and the results wouldn't be much better than the effective medium theory approach. An effective medium model can greatly simplify the original structure, whether multilayer, periodic, or both; and provide computational savings in the simulation.

16.8. Conclusion

This chapter has explained in some detail the display cube's viewing hemisphere, with a particular focus on its function as a polarizer for the invisibility cloaks that it contains. Many aspects of light physics have been explored. Corning Polarcor has been showcased because it embodies all of the pertinent design elements of the viewing hemisphere and only a few improvements are necessary.

Meanwhile, the sheer density of a 3D array that tries to achieve a continuous image with individual elements makes getting light to the center of that array one of the biggest challenges of the entire project. Polarization is at cross-purposes with getting light to the center of the display volume because polarization by definition reduces light transmittance into the cube. LED power can be increased in the core region, thereby solving the problem electrically for the color planes. For the invisibility or spacer planes, a reflection scheme using the Brewster angle can be worked out to reflect light into the interior, requiring no power and being a completely passive solution.

Finally, an overview of Effective Medium Theory as applied to reflectance and transmittance of waves on an incident surface has been presented. New mathematical problem solvers like EPACS are being used to generate metamaterial properties that, when applied by materials engineers in the lab, will lead to tangible manufactured products. There's no disconnect between math and manufacturing since materials are described by materials engineers with the same parameters that are being solved for theoretically. The language of materials is universal, so that, increasingly, tangible products will roll off the assembly line thanks to precise theoretical models coming from electrical engineers like Teng Zhao.

References

[1] Deep X-ray Lithography at http://ssls.nus.edu.sg/facility/limint.html
[2] Zhang and Wu. *Cylindrical Cloaking at Oblique Incidence With Optimized Finite Multilayer Parameters*, Optical Society of America, 2010 (online)
[3] Simon, Hilda. *The Splendor of Iridescence: Structural Colors in the Animal World*, Dodd Mead, NY, 1971
[4] Anderson, B.W. *Gem Testing*, Fourth Edition, Emerson Books, NY, 1978
[5] Corning, *Polarcor Product Information and FAQ*, 2014 (online)
[6] Barrett & Bekefi. *Electromagnetic Vibrations, Waves, and Radiation*, MIT Press, 1978
[7] Capolino, Filippo. *Theory and Phenomena of Metamaterials*, CRC Press, 2009
[8] Capolino, Filippo. *Applications of Metamaterials*, CRC Press, 2009
[9] Zhao, Teng. *Effective medium modeling*, Graduate Dissertation, Iowa State University, 2015
[10] Nicholls, Peter. *The Science in Science Fiction – From Starships to Cyborgs to Extra-terrestrials – What's Possible, What's Probable, and What the Future May Really Hold*, Knopf, New York, 1983

Chapter 17

Refresh Design

17.0. Introduction

The 3D display cube design is a level-oriented architecture, where vertical conduits spaced at regular intervals deliver data to the various levels. One can draw an analogy with a skyscraper, where the levels are floors, and the vertical conduits are elevators.

The design outlined in previous papers puts nanoscale registers at every voxel to latch the data in place. The original idea was to be able to remove the cube from its holder and have the image persist; this was accomplished with pixel memory on every level. However, this is an expensive perk in terms of real estate. In regular 2D displays, digital to analog converters take their digital input from latches ganged together for an entire row. Those latches hold only enough data for the current row scan.

In this chapter we will take the current design and extend it to make it both more efficient from the gate-layout point of view, and more transparent by minimizing signal lines in each color plane.

17.1. Current Design Review

Recall that an *optical plane* is an abstract plane oriented in the vertical direction, penetrating through all of the horizontal layers of the cube. A *color plane* is a physical layer oriented in the horizontal direction, parallel to the surface that the display cube rests on. The display cube has a preferred orientation such that its *base plane* is parallel to the color planes that reside above it.

The current design clocks color values into the cube via an optical strip which connects external memory in the cube holder to the cube's own base plane memory. From the base plane, color values are delivered to 384 ring planes stacked in the vertical dimension, given a nominal resolution of 128 voxels in the vertical direction and {r, g, b} planes for each level. The *sous-bande* adds an additional 128 values to the data payload, each value distributed identically to inner display rings of all three {r, g, b} rings.

Where the vertical conduit penetrates each ring plane, a group of 4 voxels is arranged in a *rosette*. The data for the top two voxels (called "burst A") is arranged as a {left, right} pair, which means that (384 + 128) {r, g, b | s.b.} x 2 {left, right} = 1024 eight-bit values are transferred for the entire column; and the data for the lower two voxels of the rosette (called "burst B") is arranged as a second {left, right} pair of 1024 eight-bit values. The left element and right element are transmitted in parallel.

The basic level design has 64 bus stations which span the column dimension (represented by variable *m*) for every single optical plane, with all planes spanning the row dimension (represented by variable *n*). Each bus station serves half a rosette to its left, and half a rosette to its right, for bursts A and B. The pattern is a *dual pair*: the dual part is {left, right} and the pair part is {A, B}.

17.2. Quad Patterns

A better design is also based on repeating patterns. One can strive for "elegance" but machines aren't elegant. Humans try to be elegant, and project their aesthetic of elegance to machine design, but machines have no emotion or feeling. Machines are mathematical. Delivering data in parallel, to the {left, right} pair is a powerful design element and is retained. Likewise, the 2 x 2 rosette idea is maintained, but with a new terminology: *quad*.

The basic components of a 2D display outlined in [1] can help guide the new design. For a regular 2D display the serial data is first clocked into flip flops. When the full width of the serial data has been clocked in, the outputs of those flip flops are latched into registers. This is the digital side of the display design. The outputs of those registers, arranged to suit the width of the color values, are then fed to digital-to-analog conversion circuits, which convert each registered digital value to an analog value for each pixel of the display row. Those analog outputs are gated into the 2D display with a *row scan* signal.

The display cube avoids long rows and instead focuses on quads: 2 x 2 arrays. Except for overall timing, it's not necessary to focus exclusively on resolution or even on the concept of voxels, but instead to merely develop a *cellular pattern*. A workable seed pattern is shown below.

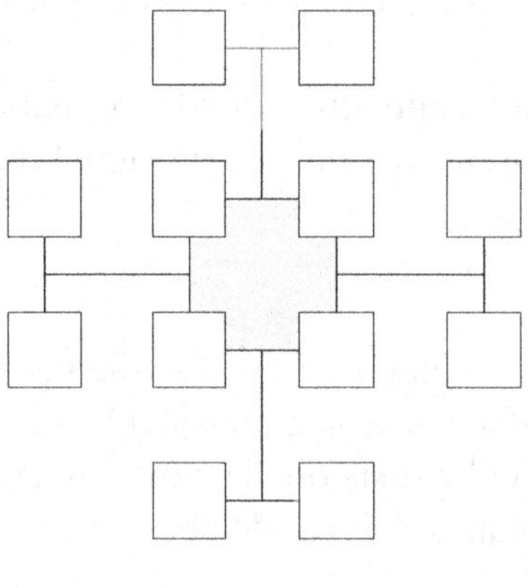

Figure 1

A central computing area shown in gray, corresponding to the central location of a vertical conduit, is surrounded by satellite areas which the central area serves. This isn't too different from the rosette idea, in that the central area is the location of the vertical conduit.

Next, the idea of *overlap* is introduced. In short, to double the "on time" for voxels, some areas of the basic pattern are overlapped address-wise and refreshed *twice*. All voxels are refreshed twice in an extended cycle of 8 refresh steps. Conceptually, the idea of overlap is applied to the pattern in Figure 1 as shown below.

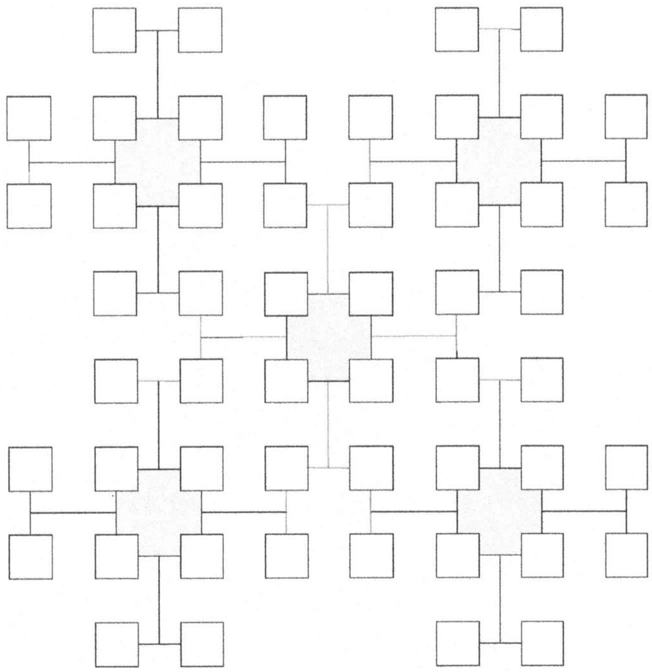

Figure 2

There two things going on here. First, the centers are staggered on rows: the odd numbered rows – for example the middle row in Figure 2 – positions the centers two elements away from the centers on even rows. Second, the *left* and *right* satellites of the second row shown above overlap with the first-row *bottom* and third-row *top* satellites. This means that all of the satellite elements around the centers (on any row) become *shared* elements. This can be seen with the small squares that have two inputs. Later it will be shown that all quads have dual inputs by taking into account *corners*.

One of the advantages of this pattern is that the vertical conduits are spaced every 4 elements in the vertical and horizontal directions. Diagonally the distance is a square root of two relation, so that, for example, for 4 elements horizontally and vertically, the diagonal direction has a length of 5.65 elements, or 4 times radical 2. The diagonal direction for an orthogonal layout like this can never equal the orthogonal lengths, even without staggering, but this isn't seen as a handicap because the eye fuses it all together when the grid is scaled to pixels/voxels.

To alleviate transparency concerns the pattern is scaled up: from a 4-center (2 x 2) with 8 satellites, to a 16-center (4 x 4) with 32 satellites, as shown next.

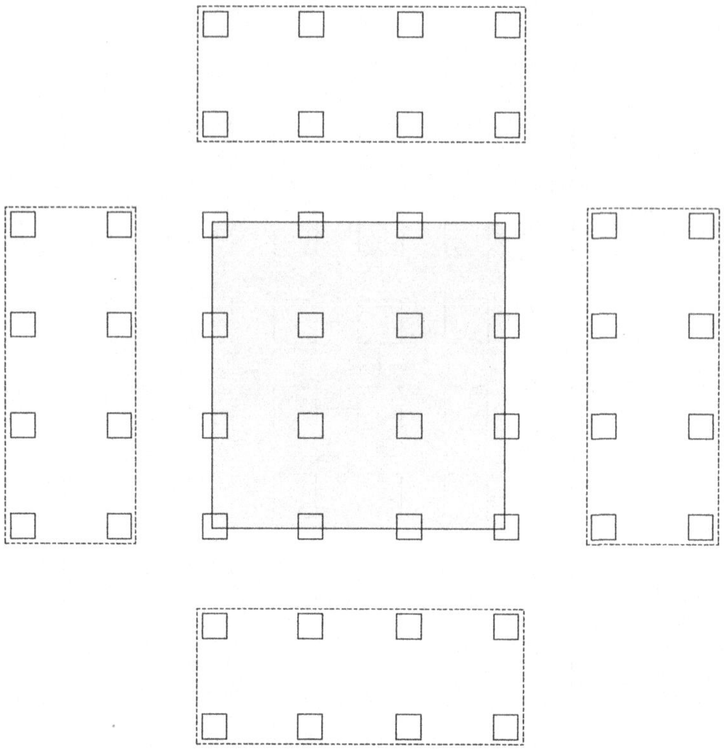

Figure 3

Notice that the cross shape doesn't fill the entire 8 x 8 array. The addressable space is $(16 + 32 = 48) < (8^2 = 64)$. The missing pixels are on the corners – and pose special problems for data lines. Overlapping the patterns gets around this problem, at the cost of some gates in the form of *multiplexers*.

The original pattern on a 4 x 4 grid is now extended to a nearly identical pattern on an 8 x 8 grid. Instead of an overlap on a single pixel, there are overlaps on groups of 4. This is shown below.

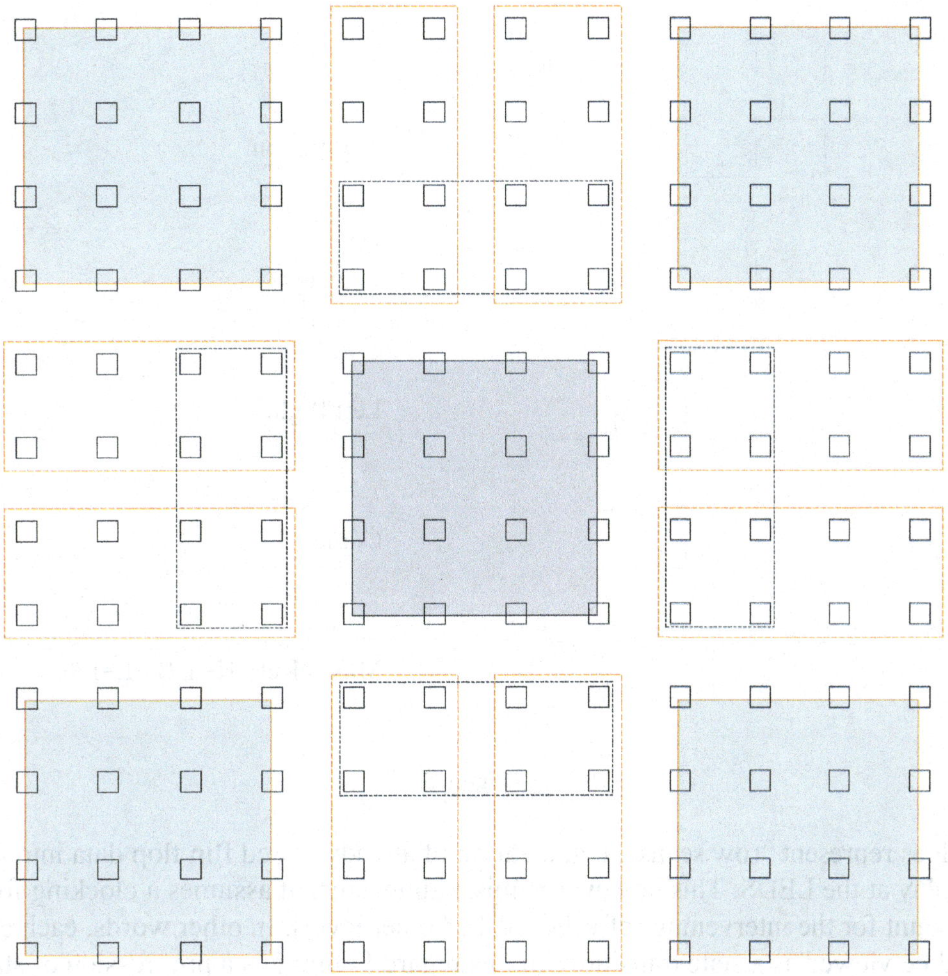

Figure 4

In this pattern, the staggered centers are shown in blue and purple with solid borders, and their satellites are shown with dashed borders. The even row centers, also called even *staggers*, are shown in blue, the odd in purple. All of this is purely for visualization. The overlap is now plainly visible on groups of four, where four voxels are shared by even and odd row centers. The general pattern in Figure 4 is the same as in Figure 2.

Since vertical conduits are assigned to "centers," the number of vertical conduits in the new design is drastically reduced. In the old "rosette" design there were, for a color plane, 64 x 64 = 4096 vertical conduit centers. In the new design there are 16 x 16 = 256 (even row centers) plus 16 x 16 = 256 (odd row centers), or 512 vertical conduit centers. This is an 8 times improvement.

17.3. Voxel Refresh Patterns

The abstract 8 x 8 voxel arrays are refreshed with a basic refresh pattern. The signal timing is shown below.

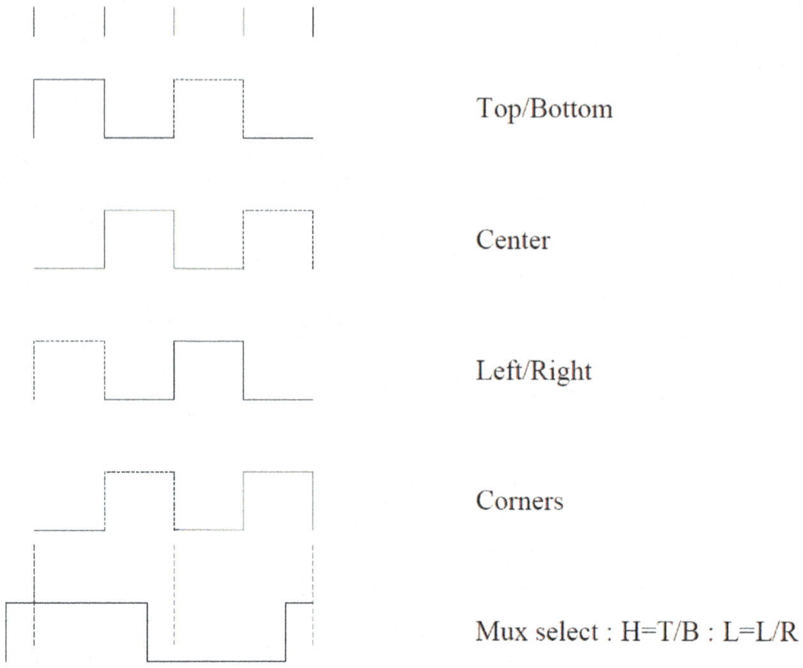

Figure 5

The positive edges represent "row scans", e.g. a gating of already stored flip flop data into the DAC and subsequent display at the LEDS. This however is misleading, since it assumes a clocking for a single level and doesn't account for the intervening refreshes of 127 other levels. In other words, each clock cycle in Figure 5 should be viewed as a state transition. In this regard Figure 5 is a progression of states.

For a cellular approach a row scan doesn't refer to a line of pixels as in a regular 2D display. Rather, these are "quad scans". The dashed signals represent the swapped refresh for row staggers. (More on this presently.) The regions in Figure 5 are identified for an 8 x 8 array below.

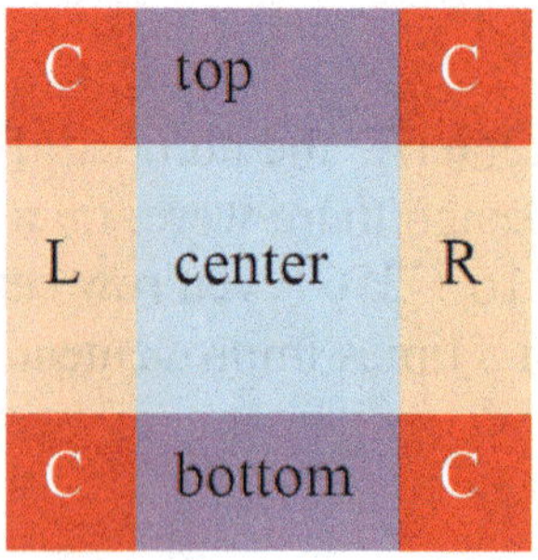

Figure 6

The corners (C) and the center form a duality since they overlap for {odd, even} staggers; a second duality encompasses T/B and L/R. The center is 4 x 4 voxels as in Figure 4, the entire grid 8 x 8. Since centers are staggered and meet at their corners, every quad region overlaps between even and odd staggers. One stagger pair is shown below, with the entire 8 x 8 region indicated on the first center only.

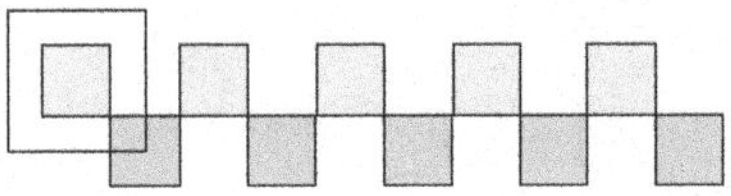

Figure 7

The sequence shown in Figure 5 is such that a *primary* refresh cycle for left/right is equivalent to a *neighbor* refresh for top/bottom. When the primary is top/bottom, the neighbor left/right regions will be refreshed. Likewise, when the primary refresh is for corners, a portion of the neighbor center will be refreshed. Also, when the center region is refreshed by the primary, those values will be used later by a corner refresh of a neighbor. Every region is refreshed (in this four-step cycle) twice. This doubles the "on time" for each voxel.

The bottom signal in Figure 5 is the MUX select. The MUX might have switching noise, so the signal transition occurs when the {top, bottom, left, right} regions are dark. Since only 12 of the 16 pixels are held in each center's registers (the corners are held by registers in neighboring centers), accessing a neighbor's center via the primary corner involves a *sneak line*, to be explained in the next section.

Imagine a regular 2D display. The entire 2D region is refreshed scanline by scanline, at 60 Hz. One reason the entire screen isn't refreshed all at once is power consumption and also the LED's *duty cycle*. Imagine trying to power a million pixels all at once! Now imagine the 3D cube. Each color plane is equivalent to a 2D screen. Activating the LEDs all at once would result in a big power spike and a possible meltdown. Instead, the display cube, like a 2D display, updates the color plane with a *rolling refresh*.

First imagine a refresh cycle (as in Figure 5) for all even staggers, followed by a cycle for all odd staggers. This is shown below.

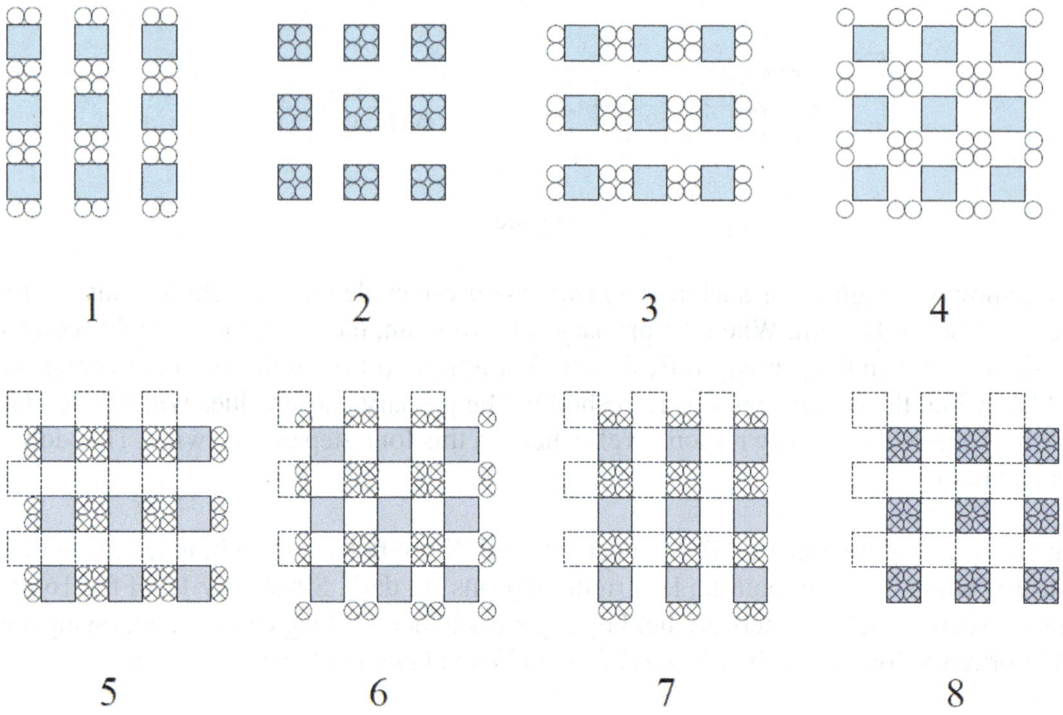

Figure 8

Steps 1 through 4 are for even row staggers, and 5 through 7 for odd row staggers. The open circles denote "first refresh." The crossed circles denote "second refresh." The *refresh order* is swapped between "pair duals" on odd staggers. In other words, {left, right} replaces {top, bottom} in the order, and vice versa. Same for centers and corners. There are two sets of duals: {top/bottom, left/right} and {center, corner}. A refresh for the entire level happens with just two refresh cycles, and so this is called a *sheet refresh.*

The sheet refresh is like refreshing half a regular 2D screen all at once, and so is not very efficient power-wise. For all even staggers, there are 128 / 8 = 16 centers per row, e.g. 16 column members. Each step activates 16 x 16 (4 quads) = 256 voxels, and there are 16 rows. This makes 256 x 16 = 4096 voxels *per step* in Figure 8. Each voxel comprises 3 color LEDs and 3 *sous-bande* LEDs. This results in activating 4096 x 6 = 24,576 LEDs all at once.

Instead of the sheet refresh a *rolling refresh* can be performed which accomplishes the same thing and is much less power hungry. In this case, only 3 rows need to be active at any one time. (This isn't to be confused with active switches at the base plane, where 2 rows are active at any one time. The base plane writes values once, but the color planes read already written values more than once.) The rolling refresh rolls through row by row, as shown below.

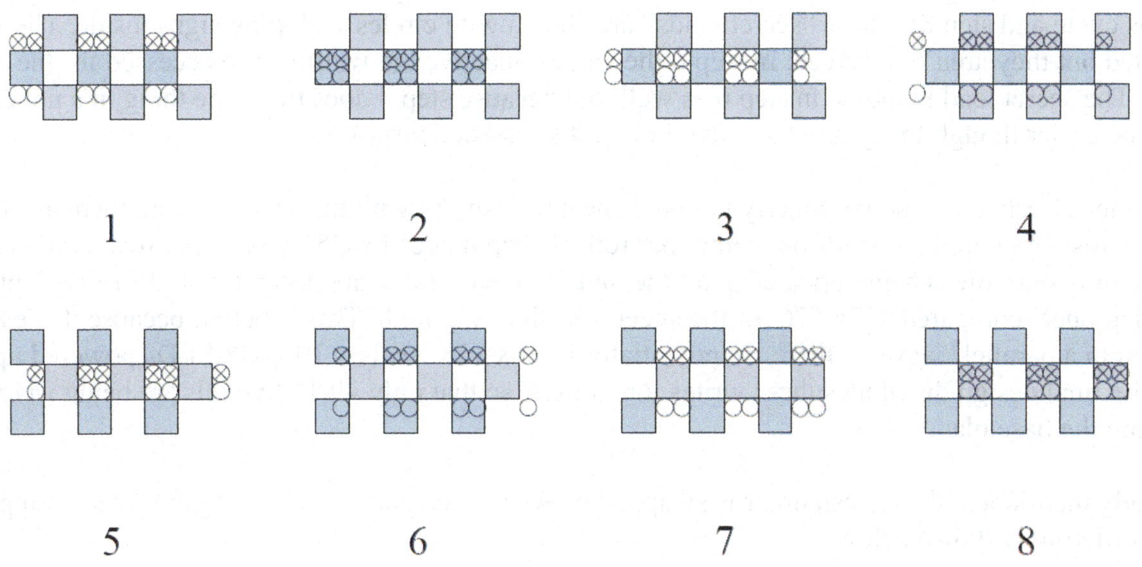

Figure 9

The even stagger in the top part of Figure 9 (steps 1 through 4) is treated in the diagram as if the rolling refresh has already passed through the rows above, and so some of the voxels are on their second refresh. The bottom part (steps 5 through 8) can be compared against the top part by reading up from step 5 to step 1, from step 6 to step 2, and so on. Again, open circles denote “first refresh” and crossed circles “second refresh.” Clearly, all voxels are updated twice, and the timing is such that the second refresh happens four steps after the first one for all locations.

Because only four values are stored at each center and processed for a refresh step before new values are written in for the next refresh step, the corner refreshes in steps 4 and 6 require one additional subtlety. The refresh as shown is conceptually correct, but doesn’t reflect the actual loads taking place. Note in step 4 that the upper corners have already been loaded from step 8 (cycling around from 8 to 1, from the previous odd stagger to the current even stagger). Once step 8 completes, the odd stagger is never written to again, so the registers in the odd stagger retain their values. But step 6 is a problem, since step 4 ends with values being written to row 3, which means the center values, which translate to the top corners for the next row, won’t be available. This problem is solved with “secret writes”. The actual writes for steps 4 through 6 look like this:

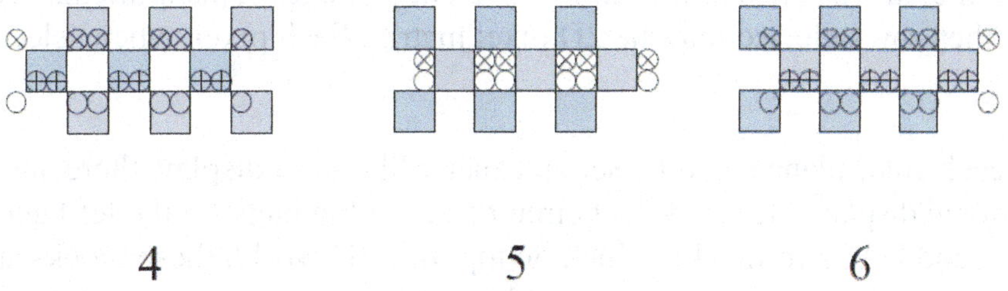

Figure 10

In step 4 the second row is reloaded for the odd stagger. This secret load takes the place of the write to the upper corners: there's no need to write to those corners because they've already been written (from the previous cycle and step 8). These "secret loads" are shown with circles with plus-signs inside. The values are loaded but they aren't displayed. In step 6 the values, inactive but waiting, are accessed for the corner refresh. The secret load is shown in step 6 as well, but because step 8 does the same thing, it's not critical (it may be easier though for symmetry, even though it's a wasted write).

The rolling refresh is less power hungry than the sheet refresh. As with the sheet refresh, there are 16 centers across. As before, 4 quads per center per refresh step makes for 256 voxels per row. However, in this case only one row is being updated at a time, not 16. Hence there are 256 x 6 = 1536 LEDs being updated at once, compared to 24,576 for the sheet refresh of Figure 8. This is better, because the levels are refreshed in a parallel "wave" – there are potentially 1536 x 128 levels = 196,608 LEDs powered up at once. The final design distributes these writes into 4 tiers, so that only 49,152 voxels are being *written to* at once from the base plane.

As already mentioned, the refresh order is swapped between even and odd row staggers. The swapped refresh ordering is shown below.

4	1	4
3	2	3
4	1	4

Even stagger

2	3	2
1	4	1
2	3	2

Odd stagger

Figure 11

In all cases there are four quads of 16 voxels per step and each quad is handled by its own processing unit. The refresh timing is governed by a clock layer which drives every center on a level. The timing is synchronous, but since timing may differ from level to level, each level operates asynchronously compared to other levels.

17.4. Color Backplanes

Getting the data out of the vertical conduit and into the color plane is the next order of business. Recall that the basic 2D display design starts with a serial register to capture the incoming serial data. From there the data is latched in parallel into a series of pre-digital-to-analog latches. This immediately cascades into the DAC circuit, and is then row-selected into the 2D pixel matrix. Each pixel, when selected, gates in the driving current.

While it's true that each color plane could be set up exactly like a 2D display, there are some reasons not to do this. In the traditional display, the row select circuit forms a big bank on the lefthand side of the pixel array, and the latches and DACs form a big block on top. In a 3D world, these blocks are walls that block viewing into the cube. The design for the cube must be sparse. This leads to a *cellular* approach.

Since data is fed to each quad from a center, then the data must be processed from the center outwards. This outward radiating scheme can be conceptualized as shown below.

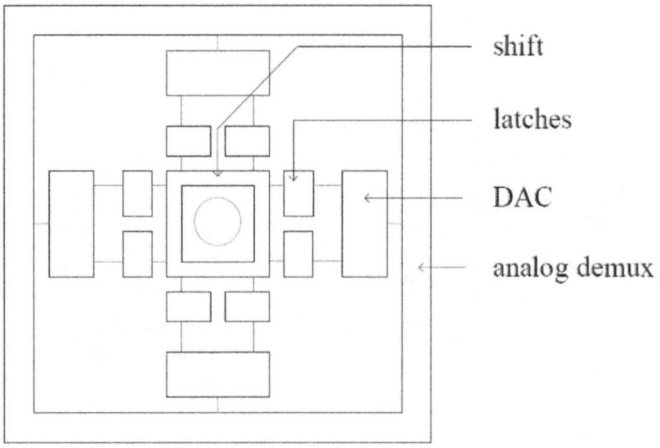

Figure 12

The data is sent up the conduit at the center (gray circle), and from there is shifted into the DQ flip flops, and working outwards, to the latches, DACs, and to an analog demultiplexer.

What is the analog demultiplexer? In a traditional 2D array, the row scan signal selects all the columns for that row. However, in the cellular approach the data can go in several directions. These directions translate into separate data lines, and the demultiplexer routes a single input to one of those separate sets of data lines. It's the equivalent of a row scan or row select in that only one of those sets is being refreshed at a time.

The conceptualized diagram of Figure 12 can be simplified further. Recall that the old "rosette" design sends {left, right} data up the vertical conduit in parallel: the design is bisymmetric because the left side is mirrored on the right.

One more conceptual leap is made by remembering that the 1:4 demux is complemented in the satellite voxels by a 2:1 mux (except for the corners which overlap a neighbor's center).*

Making the mux and demux circuits bisymmetric, and working with quads, the data line layout is shown below.

---

* The design of an analog mux/demux requires a little care, but some literature is available in [2][3][4][5][7]. Low-power CMOS design using 180 nm technology would offer the ideal tradeoff between footprint and power consumption.

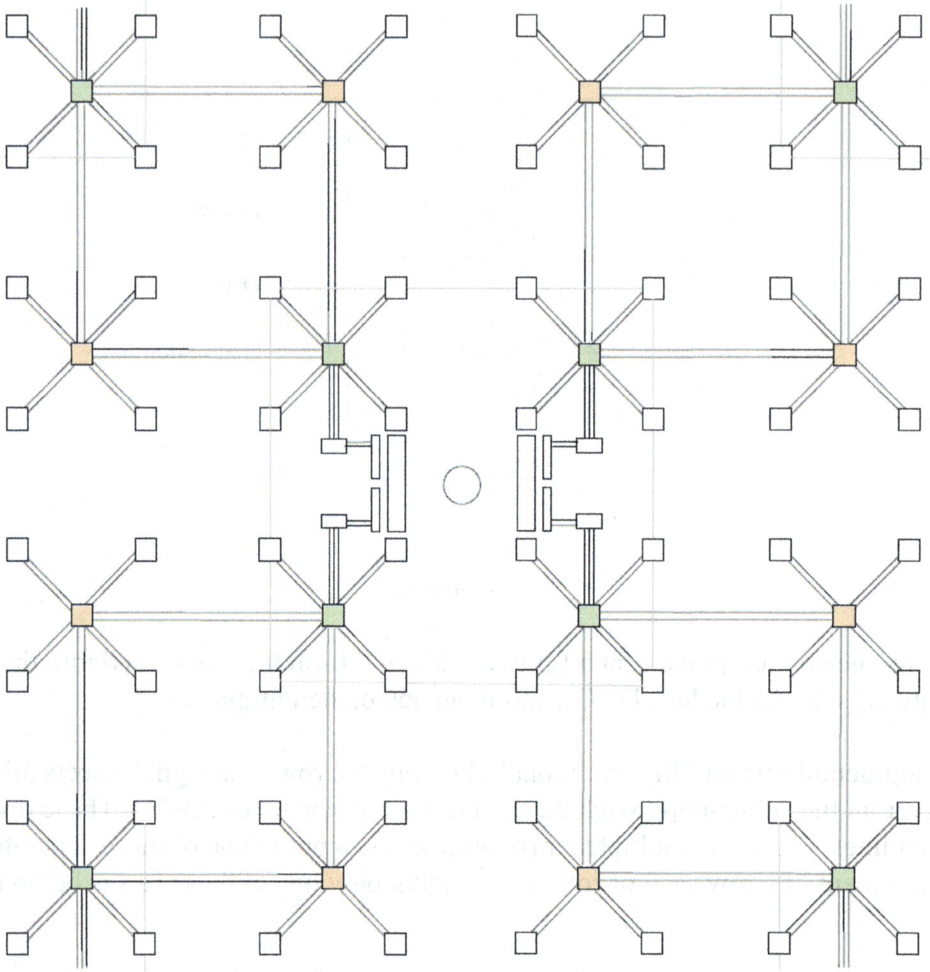

Figure 13

This circuitry resides on a *backplane*. The backplane resides on a different layer than the voxels themselves: the backplane resides just below the ring plane. As always, the backplane wiring must be as sparse as possible to keep the display cube as transparent as possible. In the old design, each circular ring has, at its center, a control island of circuitry: in essence a holding register and DAC. In the new design, digital registers are moved to the backplane. The vertical conduit penetrates each backplane at each center and makes contact with it.

Figure 13 shows how each center is split between left voxels and right voxels. The colored square pads in the center and outside of the center lie directly beneath the location of the nexus for a cluster of 4 color rings. Each *ring plane* is directly above its backplane, so the real estate of the rings themselves doesn't constrain the backplane design. Each color ring is abstractly indicated by the uncolored pads in Figure 13. Some details of the backplane design are shown below.

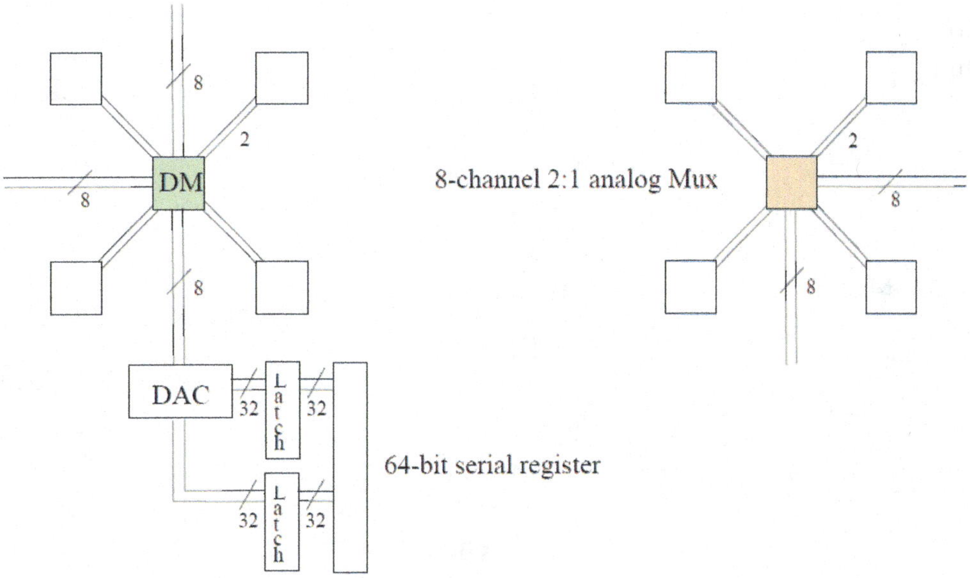

Figure 14

The reader familiar with circuit design may be wondering: if the satellite voxels can be refreshed as {top, bottom} or {left, right}, how are the two different signals selected? The answer is an *analog multiplexer*. A 2:1 multiplexer, as shown on the right in Figure 14, takes two inputs and allows one or the other to be selected as the output. This means an even row or an odd row can choose that element. In this case, the output feeds the 4 surrounding voxels. Paper [5] gives an overview of 2:1 multiplexers that feature 90nm technology.

The green box is a 1:3 analog *demultiplexer*. A demultiplexer does the opposite of a multiplexer: it takes one input and distributes it to one of three outputs (in this case). The destinations are up and out of the frame; to the left and out of the frame; and to the central quad shown by the four squares. All voxels are fed by 2 lines: the color, and the *sous-bande* value. These are analog lines.

Each small processing unit (there are four of them per center) must handle three quads: the central (or from a neighbor's perspective, corner); one of {top, bottom}; and one of {left, right}. Each quad consists of 4 voxels (which is why it's called a quad). In the digital domain, 16 bits {color, *sous-bande*} x 4 makes a 64-bit package. The {color, *sous-bande*} values for the quad are converted to 8 analog signals, which are routed via the analog demultiplexer in one of three directions. Each small processing unit must hold, then, in its flip-flop memory, 64 bits for the current quad. The processing units are labeled as left top (LT), left bottom (LB), right top (RT), and right bottom (RB). Each processing unit has its own 64-bit serial register and latch setup.

Sneak lines. The backplane design shown in Figure 13 employs a trick to accommodate the corner refreshes. Recall that only 12 voxels are written to per processing unit, omitting the 4 corner voxels. To activate a neighbor's center voxels (which overlap with a "corner"), a control line can be "snuck into" the neighbor. Since there's full symmetry, the neighbor also sneaks a control line back to reference center. This way, since each center has its own data, there's no need to route analog color signals between corners and centers. The sneak lines are labeled "4" and "6" below, corresponding to the even and odd row-stagger corner refreshes as in Figures 9 and 10.

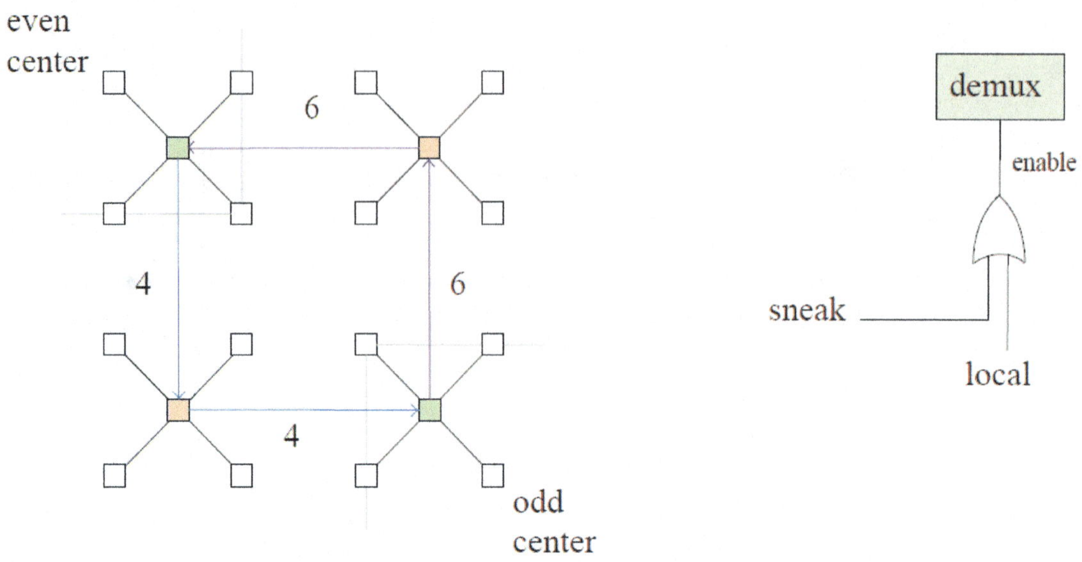

Figure 15

The pale green boxes are demuxes as shown in Figure 14, and there's a symmetric enabling of a neighbor's corner demux by way of the sneak lines, shown in blue and purple. Recall in the rolling refresh 3 rows are enabled, even though only one row is the primary row. The demux select activates corner refreshes in rows {1, 3} as part of the row {2} activity, e.g. step 4. (Again refer to Figures 9 and 10.) Remember that unless overwritten, data that sits around in flip flop registers can be used at any time.

17.5. Timing Analysis

This timing analysis establishes the dot-clock frequency which is the basis for the high-speed data transfer in the base plane. This analysis excludes base-plane timing and delays for non-data packets, which will add a minimal delay to the overall timing. Once the dot clock is established, figures for its division (slower divided frequencies) can be calculated. However these slower clock speeds are simply byproducts of the crucial high-speed frequency. The same amount of information is being transferred, whether in parallel with a slower clock, or through the optical ports with the fast clock.

One refresh step for a center involves 4 groups of 2 x 2 = 4 voxels, the selected destinations. A processing unit is designated "p.u." in the following. The destinations are encoded at the base plane as *one of* {Y, C, X, K}, for cartesian Y direction (signifying "top" for LT/RT p.u.'s, and "bottom" for LB/RB p.u.'s); center (each processing unit's home corner: each center is composed of 4 quads, as in Figure 13); cartesian X direction (signifying "left" for LT/LB, and "right" for RT/RB); and corner (with a K), being the extended diagonal to a neighbor's center. Each center holds in its registers 16 bits {color, *sous-bande*} x 4 voxels (destination) x 4 processing units = 256 bits. There are three backplanes, so 256 bits x 3 = 768 bits. This is one refresh step, e.g. *one of* {Y, C, X, K}.

Now, the cube vertical dimension is arranged as 4 tiers or groups. The write order is such that all 32 levels for the 4 groups (128 levels total) are written to for one step, before proceeding with the next step. For example, for LT, for level-groups {0, 1, 2, 3}, the sequence is as below. Each sequential step is in the vertical direction.

LT0	LT1	LT2	LT3	(step)
0	32	64	96	Y
1	33	65	97	Y
…	…	…	…	
31	63	95	127	Y
0	32	64	96	C
1	33	65	97	C
…	…	…	…	

Table 1

This means an LED can hold its value (coming from registers) for 32 vertical writes before the next refresh step comes in. This is very important, because LED persistence determines brightness. Presently it will be shown how the *hold time* can be made even longer.

Each refresh step is broken down into data packets as follows. A data packet is 64 bits. Each packet holds 16 bits {color, *sous-bande*} x 4 destination voxels. The refresh is 768 bits / 64 = 12 packets. Two p.u.'s are served by base station *n*, and two by base station *n+1*. Abstractly, 12 = 4 x 3, with 4 being (2 + 2 p.u.'s) and 3 being the color planes. The 4 processing units at each center are served in parallel. Table 1 shows the sequence for LT only.

Dot clock. The quad data for all three color planes is 64 bits x 3 color planes = 192 bits. Before being phase-modulated in the base plane the data comes all at once for the 4 processing units, so 192 x 4 p.u.'s = 768 bits. There are 128 levels, so for one refresh step we have 768 x 128 = 98,304 bits. There are 4 refresh steps at each column center, making 98,304 x 4 = 393,216 bits. This occurs for 32 column positions per link line, making 393,216 x 32 = 12,582,912 bits per optical plane. Remember that all optical planes operate in parallel via the optical strip which transfers data out of the cube's stand. A 60 Hz refresh rate for the entire cube makes 12,582,912 x 60 = 754,974,720 bits or 754.974720 MHz. Inverted, the dot period is 1.324 ns.

In 1987 this would have been an impossible number, but today it's on the low end of DDR2 RAM bandwidths, and DDR2 through 4 can be found in every modern laptop and desktop computer. But more to the point, this pixel clock is what the optical coupler must handle, and fiber optic lines typically are in the 3 GHz range, so the pixel clock is right in the ballpark.

Data is sent up each column using a slow clock of *dot* / 16 = 47.185920 MHz. Now, there are 32 levels per group. The groups are written to in parallel. Therefore, the nominal data transfer rate for a base station, per group, is 192 bits (Rx or Lx) x 32 levels per group = 6144 bits. This is for one refresh step. The amount of time to get through one refresh step is 21.192 ns x 6144 bits = 130.208 μs. The updates happen at 60 Hz and so every 1 sec / 60 Hz = 16.66 ms an update of 130.208 μs takes place. Four refresh steps takes 130.208 μs x 4 = 520.833 μs. We can make all of this jibe by seeing that for all 32 columns and all refresh cycles in an optical plane, we have 520.833 μs x 32 columns = 16.66 ms, or 1 / 60 Hz.

The "on time" for a voxel is the time for a single refresh step x 2, since the full refresh cycle updates a voxel twice. So 130.208 μs x 2 = 260.416 μs. Inverting, this gives 3840 Hz. The "on time" can be conceived as 60 Hz / 3840 Hz = 0.015625 or 1.5625% of the nominal 60 Hz refresh rate. The trick is to extend the delay between writes so that the LEDs can stay on. The way to do this is to write the first refresh step to all even staggers first (the even columns), then steps {2, 3, 4} to the even staggers; and then the first refresh step to all the odd staggers second (the odd columns), then steps {2, 3, 4} to the odd staggers for an optical plane. There are 16 even and 16 odd staggers. So 3840 Hz / 16 (the time it takes to write 16 even or odd staggers) is 240 Hz. 60 Hz / 240 Hz = 0.25 or a 25% duty cycle.

The refresh period of 60 Hz dictates how many times color values can be written to the color planes. But LED brightness depends on another clock, the pulse width modulation (PWM) clock, which is usually 4 times the refresh clock, or 240 Hz. More, in a volumetric display we don't want the LEDs too bright because they'll drown out neighboring voxels, so a duty cycle of 25% – that is, the pulse being on 25% of the time – will keep things under control. This is exactly what we have, based purely on the write times.

Why hasn't anybody built a 3D display cube already? First, these timings are for a very low resolution device: 128 voxels cubed. If the resolution is increased the refresh timing will drop precipitously, calling for new solutions that must account for power consumption, circuit density, transparency, clock speed and channel noise. The idea is to prove the concept with a toy, and if the toy works and goes into production and makes money, to use the profits to build on the initial success with a higher resolution device.

17.6. Conclusion

This chapter has presented a new level design for the 3D display cube. The new design reduces overall power consumption by reducing the number of vertical conduits. The smaller number of vertical conduits also increases transparency compared to the original design. A *cellular* approach is taken: vertical conduits penetrate each color plane in multiple locations and radiate outward.

As well, the original *color plane*, which is an abstraction of three *ring planes*, has been extended to include three *backplanes*, one for each ring plane. The signal from the backplane is only one level away from the ring plane above it. The throughway between planes is therefore very short. Compared to the original design, very little changes in the ring plane itself. However, the control island at the center of each color ring is now dispensed with; the island no longer deals with digital register logic or DACs.

A timing analysis has shown that the refresh rate for the LEDs is comparable to that for a regular 2D display. The "rolling refresh" goes a long way towards addressing LED power consumption, and yet there's no magic solution that gets rid of the power drain entirely. A certain baseline power rating will simply have to be accepted. Hopefully the LEDs can be designed so that illuminating 16 voxels x 32 per row x 32 levels x 4 groups = 65,536 LEDs at once (worst case) presents no more of a power drain than refreshing 64K worth of RAM cells for some short period of time.

Not every detail of the level design has been addressed. More details will be given in the next chapter.

References

[1] Ying, Wang. *Silicon Backplane Design for OLED-on-Silicon Microdisplay*, Master's thesis, Nanyang Technological University, 2011, online
[2] Wang, Ruixue. *A Low Power 8 to 1 Analog Multiplexer for Bio-signal Acquisition System with A Function of Amplification*, Master's thesis, Carleton University, 2016, online
[3] Philips Semiconductors, *74HC/HCT4351 8-channel analog multiplexer/demultiplexer with latch*, data sheet, 1990, online
[4] Analog Devices, *Analog Switches and Multiplexers Basics*, MT-088 Tutorial, 2008, online

[5] Gupta, Arora, Singh. *Simulation and Analysis of 2:1 Multiplexer Circuits at 90nm Technology*, IJMER, vol 1 issue 2, 2011, online

[6] Carpenter, Aaron. *The Design and Use of High-Speed Transmission Line Links for Global On-Chip Communication*, PhD Thesis, University of Rochester, 2012, online

[7] Chen, Wai-Kai. (Ed.) *Logic Design*, CRC Press, Principles and Applications in Engineering Series, 2003

Chapter 18
Base Plane Logic

18.0. Introduction
This chapter looks more closely at the logic and silicon design for the cube's base plane.

18.1. A Crash Course in Current Mode Logic
The *base plane* positioned at the bottom of the cube operates at very high speed, and not all logic families will support such a high speed. Current mode logic (CML) uses differential signals to achieve stability. The signal is divided into a positive and base reference voltage on the negative side (usually ground); or in a mirrored fashion with a negative voltage and a base reference voltage (again ground), allowing data to be transmitted at speeds between hundreds of megabits to tens of gigabits per second. The transmission is point to point. Point to point means signals only travel to and from end points, quite different from traditional busses that use high impedance to share data between multiple devices. A transmitter is linked to a receiver, and the *link line* is terminated at the receiver.

CML is used for high-speed connections between modules, connections between modules on a chip (no signal leaves the chip surface), and for frequency dividers and multipliers. Anything high-speed. The NEC SX-3 supercomputer used CML for high-performance, high-density computing modules. The SX-3/44 was the fastest computer in the world between 1992 and 1993, with a clock cycle of 2.9 nanoseconds. Does that mean the display cube, like the SX-3/44, will cost millions of dollars? Not at all. Differential logic is only used on the cube's base plane.

The basic block diagram for a transmitter, and its CML circuit, is shown below. (See also [1].)

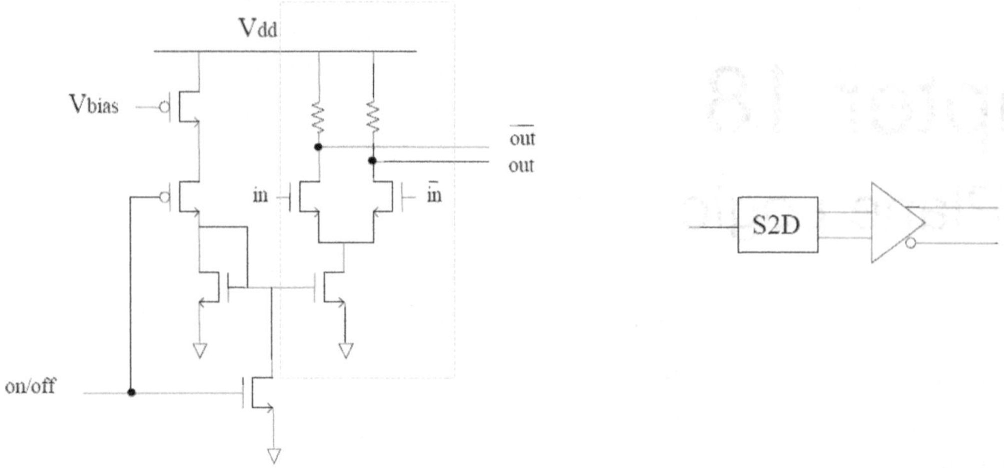

Figure 1

The output driver, represented on the right by a right-facing triangle, is shown on the left as a CML circuit diagram, with the output driver proper inside of the dashed box. The inputs and outputs are doubled, as with all differential circuits. CML is an analog solution, though the analog signals transmit digital data. In this sense perhaps any logic family, such as TTL, is analog, but the high-speed nature of CML emphasizes the point. S2D in Figure 1 stands for "serial to differential." Serial means only one line, and a circuit can convert from serial to differential, and another circuit from D2S.

A transistor can be represented as a collector (in all cases in Figure 1 the top leg of the transistor), a base (denoted with double lines and an input), and an emitter (shown with arrows). (Source, gate, and drain are similar terms for MOSFET.)

In *active* mode, two junctions are differently *biased.* The emitter-base junction can be *forward biased*, and the collector-base junction *reverse biased.* Current in this case flows between the emitter and collector. Think of the current as flowing into the base in the forward direction; and flowing out of the base at the collector. The collector doesn't let current into the base. The amount of current flow is proportional to the base current. This makes the transistor an amplifier.

In *cutoff* mode, the collector-base junction is *reverse* biased, and same for the emitter-base junction. Now the *emitter* doesn't allow current to flow into the base. When the base-emitter voltage is low, current doesn't flow from collector to emitter. The device is completely switched *off.* (Although in reality there's always a little *leakage current.*)

In *saturation* mode, the collector-base junction is *forward* biased, and same for the emitter-base junction. Now both ends allow current to flow into the base. When the base-emitter voltage is high, current flows freely from collector to emitter. The device is fully switched *on.*

Figure 1 uses bias to drive the circuit closer to or further from saturation, depending on the circuit signals. The bias of CML logic is always set with reference to a *bias reference generator*. The reference generator is designed to maintain constant bias with power supply and temperature variation, and is connected to a voltage bias input. Below, a regular CMOS inverter is shown on the left. The driver circuit from Figure 1 is shown on the right for comparison.

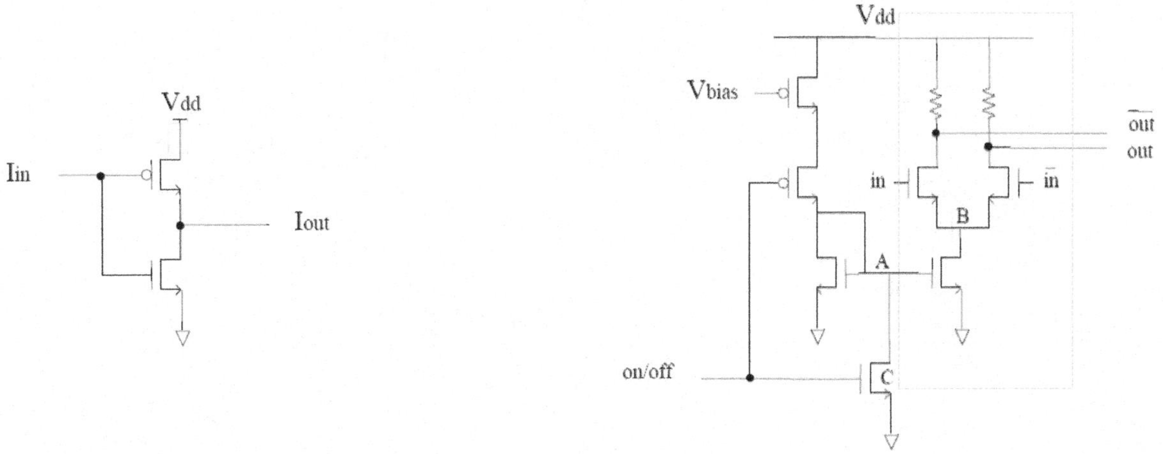

Figure 2

The top part of the inverter, e.g., the topmost transistor, is a PMOS current source. PMOS transistors show the "open dot" on the base and become conductive with a negative voltage. NMOS transistors have no dot and become conductive with a positive voltage. PMOS and NMOS are said to be logical "duals" of each other. Underneath it all, both flavors fall under the category of *metal-oxide semiconductors* (MOS).

When Iin is *high*, the current from the PMOS transistor (the top part of the diagram) will flow through that branch, leaving no current to go to Iout. The PMOS becomes "non-conductive". Meanwhile, the NMOS becomes conductive and the output is a low voltage (since the NMOS is tied to ground). If Iin is *low*, the PMOS is conductive, the NMOS is non-conductive, and the output is a high voltage. So this circuit inverts the input.

On the right, the on/off input is fed into a similar inverter circuit, except at point "A" there's a "wired and" into a bias reference generator. The reference generator is feeding into the CML amplifier at B. B is a "series connection" and performs a logical "OR" on "in". The NMOS between A and B is part of the reference generator, but notice that the reference generator itself is being drained by C.

A reference generator circuit is shown below on the left. [2] Again, the driver circuit of Figure 1 is shown for comparison to the right.

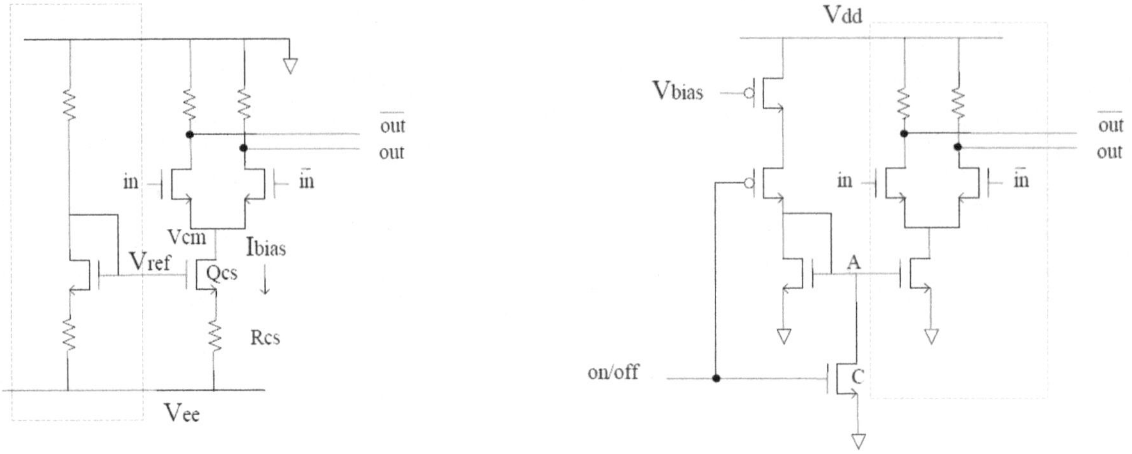

Figure 3

There are many similarities because the output drivers are the same. One big difference is that the circuit on the left is referenced to ground on top, and power on the bottom – the exact opposite of the figure on the right. However Vee is actually a negative voltage, about –5.2V, so the circuits are equivalent because they go from a reference voltage (on top) to a more negative voltage (on the bottom). (Vdd is a positive voltage: about 2V.) The reference generator part of the circuit on the left is shown in the dashed box.

In short, the average value of Vcm remains constant because changes in the power supply (+/– 0.5V) are absorbed by the base-collector junction of Qcs. The current mirror on the inputs (where label Vcm resides) sets the bias (Ibias) for the inputs. With constant bias, as Vdd moves by +/– 10%, the voltage drop across Rcs remains constant. All of this then keeps the logic from being affected by changes in the power supply.

The driver circuit on the right side of Figure 3, where A plays the same role as Vref, is part gate and part reference generator. An "on" signal (high) saturates C and makes A conductive – equivalently Vref in the circuit on the left side is conductive. Bias correction is applied. An "off" signal (low) turns the reference off and nothing is amplified in the amplifier. So much for the transmitter.

The receiver circuit is as follows [1]:

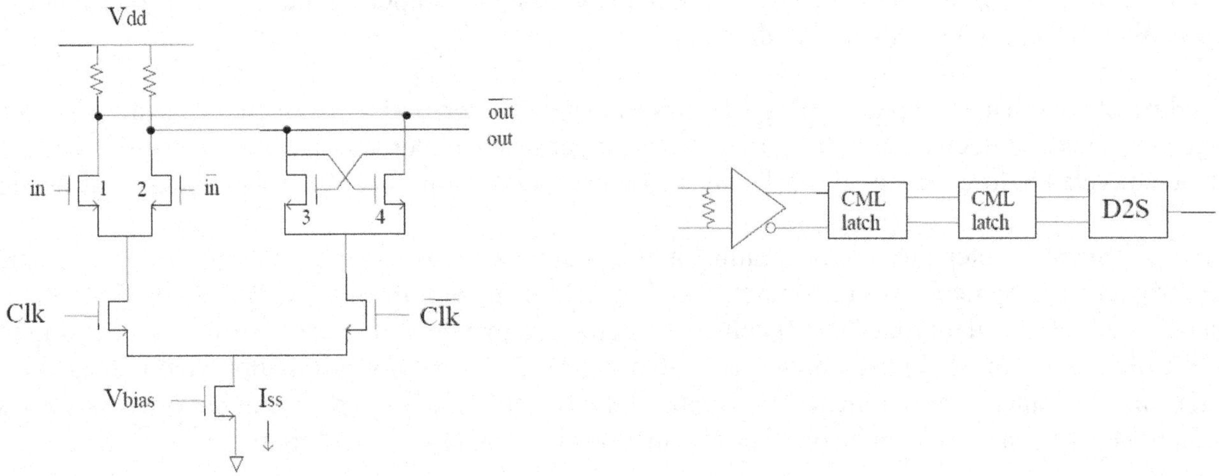

Figure 4

The explanation given in [3, 4] is as follows. This circuit is a CML sampler and latch. The transistors labeled {1, 2} are an input tracking pair, or sampler, that senses and tracks data variation. Again, bias is coming in through the current mirror at the bottom, but this time the switch is a clock signal (Clk). The circuit therefore acts as a *data follower*. The transistors labeled {3, 4} are a cross-coupled regenerative pair, used as a latch. Note that this circuit requires a low-skew clock. When the clock is *high*, V_{out} tracks V_{in}. When the clock is *low*, the latch is enabled, which stores the logic state at the output. As usual, there's a voltage-bias input to keep the current stable.*

On the right, the CML latch is shown in context. It's understood that each differential length or run should be terminated before reaching the amp, to avoid signal reflections. That's what the resistor is for. The entire receiver circuit is described as a "differential amplifier plus current-mode-latched sampler." [1] Heydari [4] improves on this sampler circuit by adding a few more pull up resistors and distributing the bias, e.g. the reference voltage. "D2S" is a "differential to serial" converter – the end of the line for the serialized differential transmission.

18.2. Base Plane Design

The *base plane* receives data from the cube holder over a high-speed optical link. The data is routed to the vertical conduits. The base plane is best served by the vocabulary of the *system on a chip* (SoC) and the related *network on a chip* (NoC) design ethic. Ideally, the base plane should be manufactured from a single die.

The dissertation by Reehal is a good overview of NoC architecture [6]. One of the main concerns of putting many components on a chip is power consumption. The display cube is a repetitive architecture and the

---

* Bias is determined by what the circuit is supposed to do. Recall that a transistor can be in saturation. If the inputs are forward biased this results in a free flow of current. A circuit can be designed at the board level so that less current flows through a transistor, e.g. the transistor is "biased to a more negative voltage". Full saturation is necessary for maximum speed but is sometimes levelled off because full saturation also generates heat. Sometimes a transistor is designed to absorb current from other transistors and needs "headroom."

elements are multiplied many times, which means various tricks must be used in order to conserve energy. The cube is an *interconnection network.* The base plane design is important because the color planes are useless unless data can be delivered to them.

Reehal points out that as chip technology continues to get smaller, and as more elements are packed onto a single chip, bus architectures will be harder to design because smaller and faster also means a smaller distance traveled before the signal breaks up, so a more modular approach to network design is needed.

As more elements contend for the bus, timing and layout become a challenge. A NoC design dynamically delivers packets to switches point-to-point, avoiding the bus paradigm entirely. NoC logic finds a path for data between Intellectual Property (IP) cells: nameless cells that perform some essential computing task. An IP cell from a NoC design standpoint is treated as a black box: it doesn't matter what it does. All that matters are the routes between them. The layout of the IP cells is its own problem, but obviously constrains the design of the links between boxes. These tradeoffs are what NoC is all about.

The base plane of the display cube doesn't need a full NoC architecture: available patterns can be simplified. The data for the color planes is delivered serially, but that data arrives in stages. First, there's the high-speed serial link. Then the data is delivered in parallel to a set of vertical conduits: this parallelism allows the clock speed to be reduced, which minimizes problems such as clock skew that are associated with high-speed data.

As the data comes out of memory it's serialized to a high-speed data clock, and the main concern is "clock recovery" on the receiver side, e.g., inside of the cube on the base plane. (See Chapter 5.) After the clock is recovered, the data packets can be decoded by the "control logic" in the base plane.

High speed data lines are inherently noisy. The differential scheme in [8] reduces noise by balancing the transistor currents with two inputs: the positive input and its inverse, e.g. a "differential". The *landing pad* for the incoming serial stream, situated on the cube side of the optical strip, is where the clock signal is recovered, as well as the digital serial stream, and the whole thing fits within a footprint of 300 μm x 500 μm, which is about the footprint of a quad. This "3 by 5" circuit by Zhuo is described in Chapter 6. The circuit in [8] doesn't include clock recovery, and its footprint is 800 μm x 200 μm. Whereas Zhuo treats the whole system but leaves out details of signal quality, [8] goes into detail about the signal quality and leaves out clock recovery. In any case, the required circuits are the right size for the cube base plane.

In the old design the vertical conduits were placed at every other voxel and an optical strip with 128 landing pads or receivers, one for each abstract *optical plane*, was considered acceptable. This number can be reduced to 128 / 4 = 32 optical ports, as explained in Chapter 17. Each port takes the serial data and converts it into four separate signals, by using a 4-phase multiplying delay-locked loop (Chapter 7).

The signals are named for the processing units residing in the color planes. Recall that there are four processing units: left top (LT), left bottom (LB), right top (RT), and right bottom (RB). At each of the 32 optical ports, four phased differential link lines carry signals to alternating centers, left and right. If the landing pads are oriented along the x-axis, then each *link line* travels in the y-axis direction. (Most of the following diagrams will adopt the convention of y-axis vertical.) These directions will be labeled *m* and *n* respectively.

The table below is to be read for two adjacent vertical columns, at *m* and *m*+1. For each row *n*, two lines on each link will always be inactive, the other two active. The active lines are the reverse for the next column over, for the same row *n*. For the *next* row *n*, the identities are likewise swapped: if LT/LB were active, then RT/RB become active on the next row. This may seem unnecessarily complex, but the aim is mathematical symmetry, which can only be achieved for the cube setup with some redundancy. This arrangement corresponds to the even and odd *staggers* of Chapter 17.

m (col)					m+1			
active		inactive			active		inactive	
LT	n (row)	RT	n		RT	n	LT	n
LB		RB			RB		LB	
RT	n+1	LT	n+1		LT	n+1	RT	n+1
RB		LB			LB		RB	

Table 1

The link-line layout in a rough manner is as follows.

Figure 5

The serial link lines are shown in gold. Each link line has *drops* or *taps* that alternate between *even staggers* (with centers on {*m*: 0, 2, 4, …}) and *odd staggers* (centers on {*m*: 1, 3, 5, …}). Remember *m* goes vertically. There are 32 optical ports horizontally, from which extend 32 link lines, each of which addresses base stations along its length, to either side, left and right. The optical strip is shown on the upper edge in red, simply to reinforce the idea that the serial data is coming from an external source. The larger boxes shown as red outlines around the centers represent the complete 8 x 8 voxel array, and it's clear that the arrays overlap. Each base station acts as a high-speed switch.

Each switch is composed of a high-speed *front end* and a lower-speed *back end*. The back end is responsible for transferring data to the vertical conduit. The fastest circuits are closest to the link lines, and the slower circuits are farthest, not only because of the physical constraints, but because this graduated placement helps to avoid noise coupling into sensitive high-speed circuits.

The high-speed part of the switch, which is really the link line itself, is shown below.

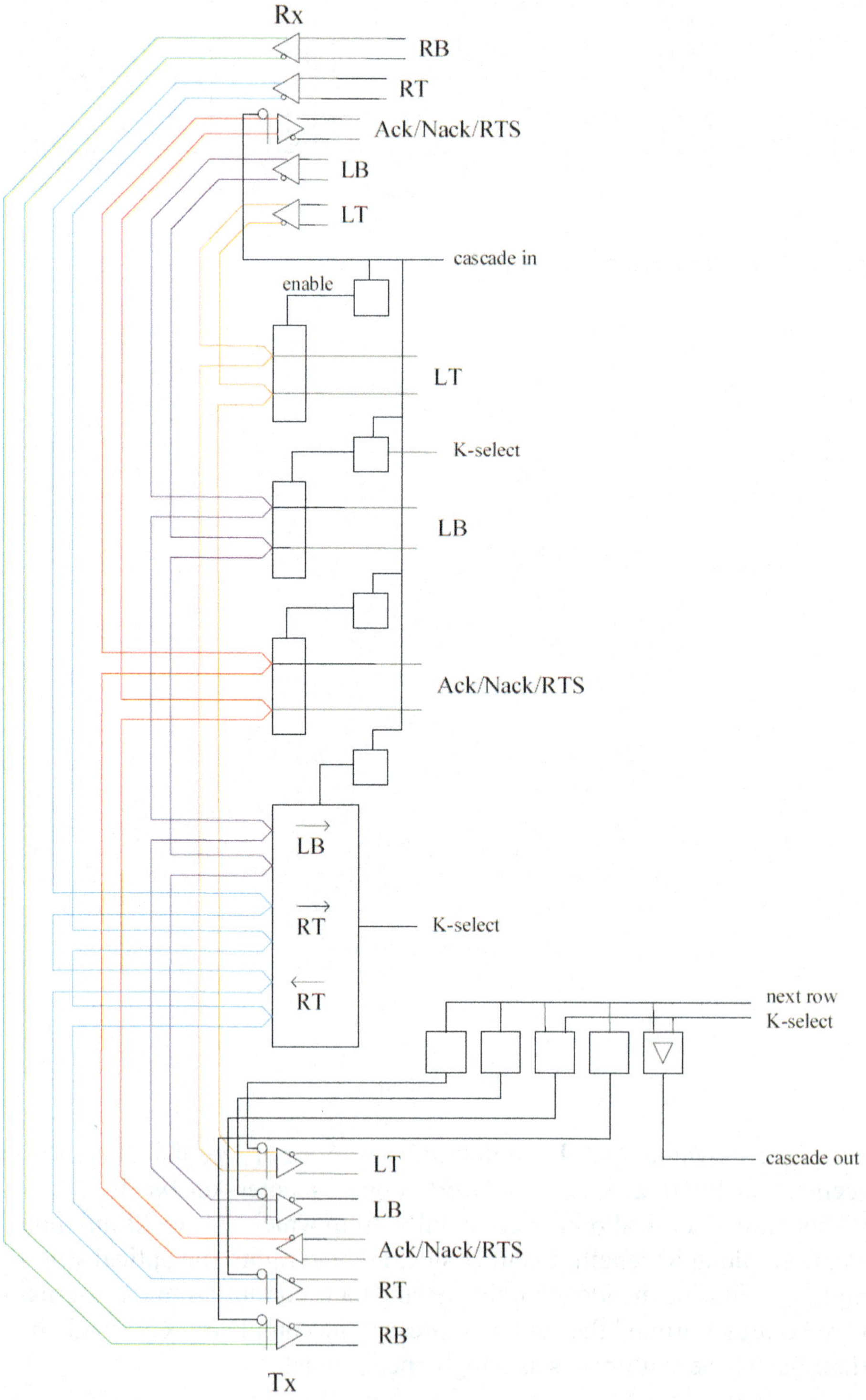

Figure 6

Note that this diagram is for the active LT/LB pair. The differential lines are color-coded for readability. The amplifiers shown with triangles also signify the direction of the data flow. Most of the lines are flowing in the same direction which makes the link *unidirectional*. However, a single Ack/Nack line communicates with the landing pad at the optical port, which makes this design a point-to-point *bidirectional* design. The transmitters are gated, shown with a circle and vertical line at the bottom of Figure 6, and those gates are on/off switches driven by logic that will become clear presently.

The diagram shows how the taps can be turned on and off with an enable line. This is crucial when a row is no longer active and the data is being routed to another center. When the taps are enabled, the signals should be able to pass into the base station with no distortion or signal loss. When disabled, very little energy should be able to pass into the switch in either direction. The signals bypass the switch in this case and continue on, just as if there were no taps at all.

The taps create some complexities involving *impedance*. The stubs leading to the taps, branching off the main point-to-point lines, change the loaded impedance of those lines. As capacitance is added to the bus in the form of devices and interconnects, the impedance is lowered. Impedance mismatch results in reflections. Luckily there are only three impedance states for the active lines: receiver to current center; receiver to current center plus k-center re-route (used for corner writes); and receiver to transmitter without taps. These three states can be corrected for with a resistor network. (See also [13].)

Everything in Figure 6 is shown on one PCB layer, but in reality this station will occupy multiple PCB layers. In that regard, Figure 6 is only a sketch, but it's something a PCB designer can work with. Note that the cascade in/out lines are serial, not differential. Everything else coming in, at this point, is differential. The next diagram follows the signals into the interior of the switch aligned with the current base station or center.

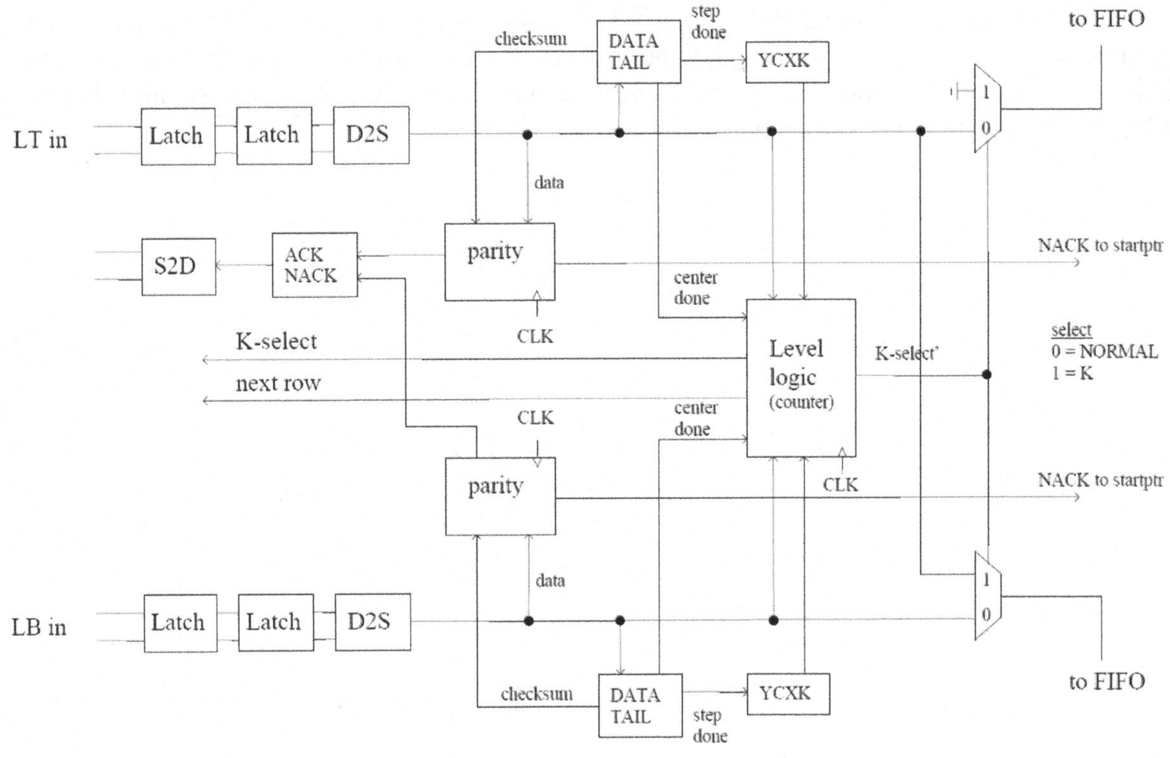

Figure 7

Figure 7 can be thought of as being "to the right" of Figure 6. This is transitional logic, the first part of the serial logic, but all operating at high speed. The decision is made to include basic parity checking on the data. For corner writes, LT is routed to LB, and as Figure 6 shows, LB is pre-routed to RT of the next row. This means the LB and RB lines need to be disabled on the primary center when they're being routed to neighboring centers. The clock signal plays a big part in Figure 7. The clock must be synced to the CML tracking in the latch receiver circuit (itself driven by a clock, see Figure 4).

The "level logic" and "data tail" boxes can be thought of as packet *header* and *tail* decoders. The basic paradigm is for a *pull*. Each switch along a link line is enabled in turn via cascade signals. The *first* switch for the first column on each link line is enabled via the "cascade in" by the landing pad logic at the optical port. But once that first switch is enabled, the enabling proceeds internally: each switch enables the next switch, from switch to switch in a cascade.

Once a switch is enabled, it sends a "request to send" (RTS) signal through the ACK/NACK line, which is the *pull* signal, and then acknowledges (via ACK/NACK) data packets that are sent in response. RTS *high* means that the switch is ready to receive data. The transmitter (at the optical port) then sends the first data packets. The packets are in NRZ format (non-return-to-zero) with a parity bit, and start and stop bits. The packet is accepted when the parity check succeeds and the stop bit has also been received; the switch sends an ACK. If there's an error, the switch sends a NACK and the transmitter retransmits from its backup buffer.

Retries can throw the timing way off, unless handled. This means that the *cube holder* has to have bidirectional optical ports, to receive data as well as send. For example, the data could be corrupt on the cube side, in which case the holder is told to "resend" the last packet, or in the worst case, to resend the entire optical plane.

In the NoC world, data and control packets are called *flits*, short for *flow control units*. Once deserialized, the data can be broken into bytes, for example, and the switch then starts dealing with *physical units*, called *phits* for short. The external memory is neatly arranged so that a series of packets are serialized into five different kinds of flits. These are shown below.

4-Phase Leader

Sync	(lock leader)

Data header (Gated through 4-Phase)

Type	Level in vertical dimension	Primary center address in base/color plane	Y/C/X/K

Data packet (immediately follows data header)

Start bit	Fixed length data (64 bits)

Data tail

Type	Stop bit	Checksum

Level tail

Type	Done for optical plane	Done for center, all steps	Done for step

Figure 8

The packet consumption proceeds from left to right: that is, the Type/Sync is consumed first, followed by the next chunk and so on. The flits have different lengths. The data packet is 64 bits in length. The data is split into 4 serial streams (see the 4-phase circuit in Chapter 7), one stream each for the left top, the left bottom, the right top, and right bottom processing units. The data is delivered to each ring plane {r, g, b} for a level and the levels are loaded in sequence. The selected quadrants for the current refresh step are:

Y/C/X/K cartesian Y, center C, cartesian X, or corner K

See Chapter 17 for details. The delivery destinations are complicated by the corner writes because the corners belong to other centers. The refresh scheme solves this problem by writing ahead to the next row for the lower corners. This is also solved in the base plane by "K-selecting" the next row, so that data is processed using default center logic in the next row. This "K re-route" can be seen in Figures 6 and 7. Specifically, the row below the *primary row* is accessed only for the writes to lower corners. Then, on the next round of refreshes, the next row becomes the primary. Values retained from the previous round for the upper corners are easily accessed (they don't need to be rewritten), and only lower corner values are written. This sequence of steps happens repeatedly for pairs of rows until all 32 rows of the column are refreshed.

The *back end logic* is shown next. This diagram only applies to the LT data that's coming in. A similar circuit for LB data exists, as well as mirrored circuits for RT and RB on the *m+1* link line. The *modus operandi* for this circuit comes from [5].

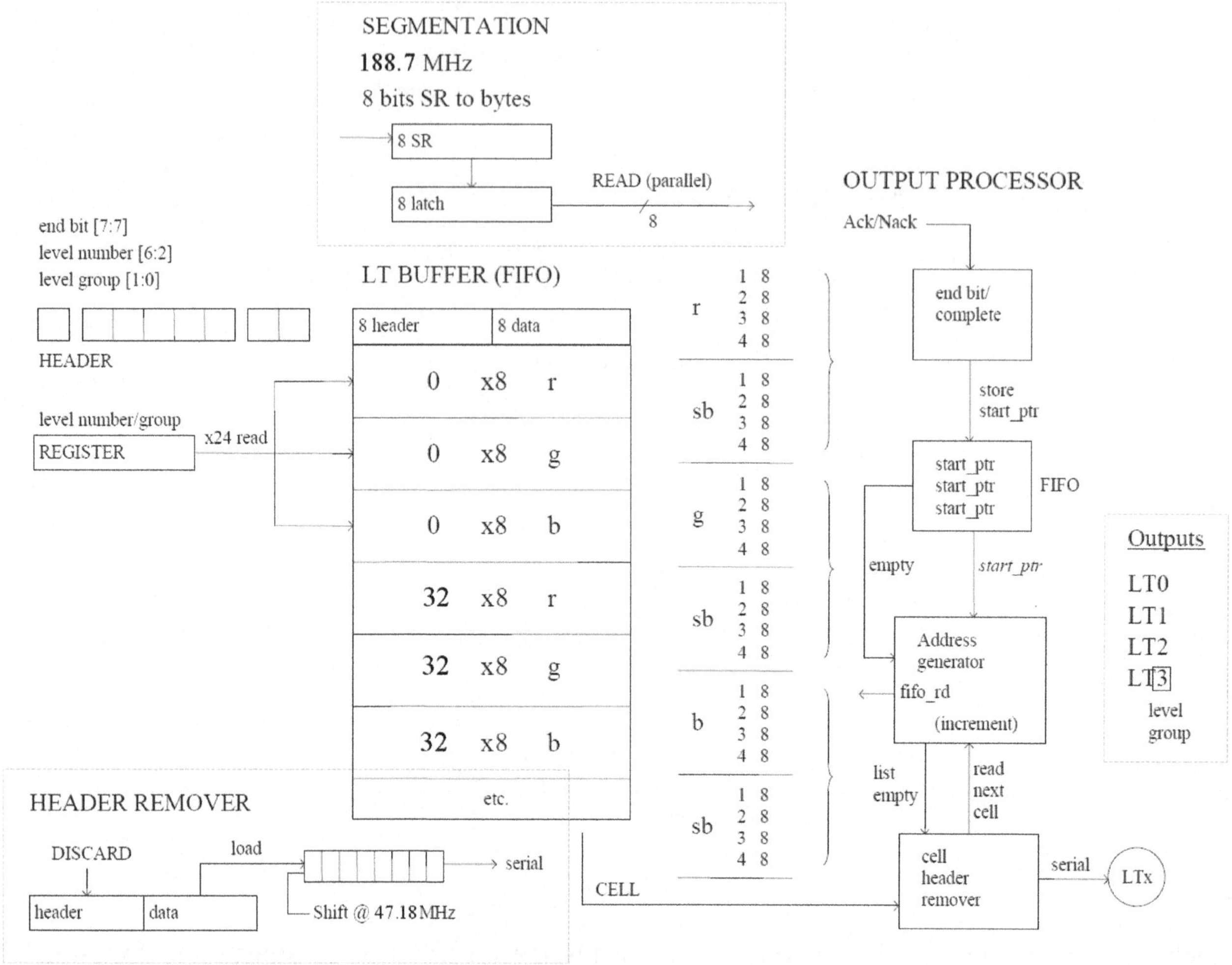

Figure 9

The output at the lower righthand corner goes straight up the vertical conduit. Each original packet is broken up into 64 / 8 = 8 *cells*, each 16 bits wide. The original packet composed of {color, *sous-bande*} data x 4 voxels is deserialized, but only temporarily. 8 bits are for a header, and 8 bits are for the original data, now in the form of 8-bit *phits*. The header format is shown to the left in Figure 9. The phit headers are not to be confused with the flit header in Figure 8. The flit header and tail packets are deserialized upstream, to keep the counting logic in Figures 7 and 9 in sync, and they serve to track the levels and groups.

The "end bit" in Figure 9 tells the back-end logic that the packet load is complete. The cells are stored in a FIFO, and as well a group of start pointers are stored in a separate FIFO, one start pointer per 64-bit data packet. The column to the right of the FIFO shows how the data is stored, 64 bits at a time ("sb" stands for "sous-bande".) The FIFO shows two groups identified by the current level in each group {0, 32, …}. The complete storage sequence is shown below.

LT0	LT1	LT2	LT3	(step)
0	32	64	96	Y
1	33	65	97	Y
...	...	...	...	
31	63	95	127	Y
0	32	64	96	C
1	33	65	97	C
...	...	...	...	

Table 2

Table 2 is to be read left to right, and top to bottom. It shows the pattern for LT (the top left processing unit in the backplane). All four level groups are being written to in parallel. There are 3 x 4 = 12 other lines for the other 3 processing units, making 16 data lines total for a vertical conduit. Each group has cascade control lines as well.

Paper [10] describes the address generation for the FIFO shown in Figure 9. One nice addition is that, since the data is stored in the FIFO before the parity check is made (see Figure 7), if the parity check fails, the start pointer can simply not be stored in the start_ptr table, and the bad data will be overwritten by the next incoming packet. The FIFO is in essence a circular buffer. Meanwhile, with no start_ptr to be retrieved, the bad data, even if it hasn't been overwritten yet, won't be sent up the vertical conduit. This is a *handshake circuit* between the front-end (Figure 7) and the back-end (Figure 9).

The FIFO is fast dual-port RAM. The FIFO is a project in itself and will require a large portion of the nano-footprint of each base-plane switch. The following diagram shows how the LT FIFO actually consists of four separate FIFOs so that the data can be read out in parallel for each level group.

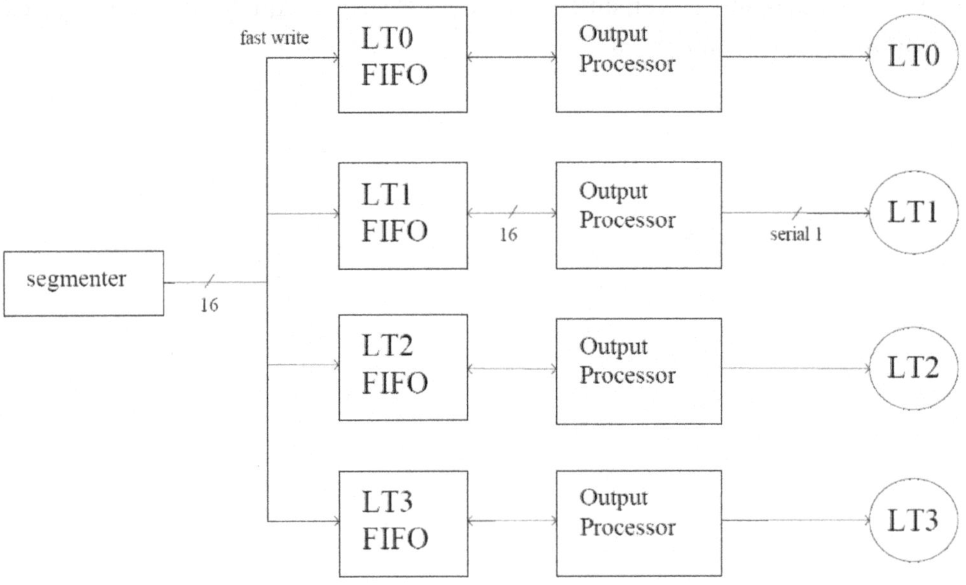

Figure 10

This is true parallelism and allows the clock to be slowed down to 47.18 MHz for the data being sent up the vertical conduit. Recall from Chapter 17 that the dot clock is 755 MHz, and this is reduced by a factor of 4 to 188.75 MHz, because of the 4-phase parallel decoding. Each extracted data stream corresponds to LT, LB, RT, and RB.

Recall that the multiplying delay-locked loop (MDLL, see Chapter 7) creates four streams clocked at 188.75 Mhz, but phase-shifted. Each stream is sent down its own differential link line. This means that the color data coming from the cube holder is arranged as a stream of {LT, LB, RT, RB} bytes, in sequence, which are then split into four streams by the MDLL. Once the data is split into streams, each switch gets a steady stream of data at the lower clock speed.

The data packet is 64 bits: 4 streams x 8 bits x 2 {color, *sous-bande*}. Once the interleave is separated into separate streams, each stream accumulates 4 voxels, being the current refresh step, and these are reassembled as a new data packet for the processing unit (LT, LB, RT, or RB). Three color planes consume the data from 3 of these reconstructed 64-bit packets for each link line. For 32 levels there are 3 x 32 = 96 packets per link line. This is the data for one refresh step. The transfer from holder to cube is shown below.

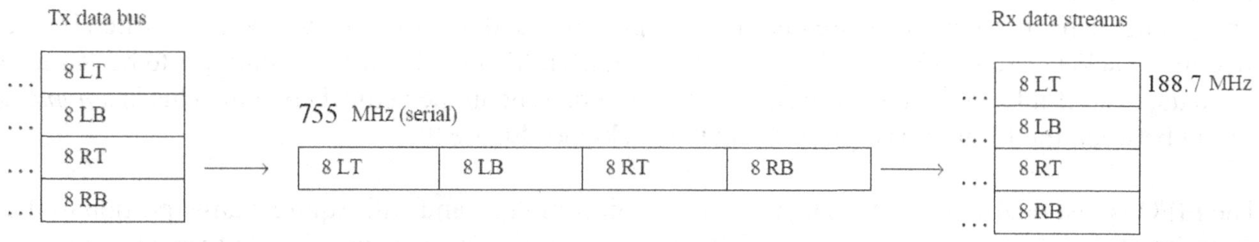

Figure 11

The next diagram shows how signals are being routed. This mux setup is the "key" to the routing of the processing unit data in Figures 6 and 7. Remember that corner writes require access to the next row, which is partly what this routing scheme is all about.

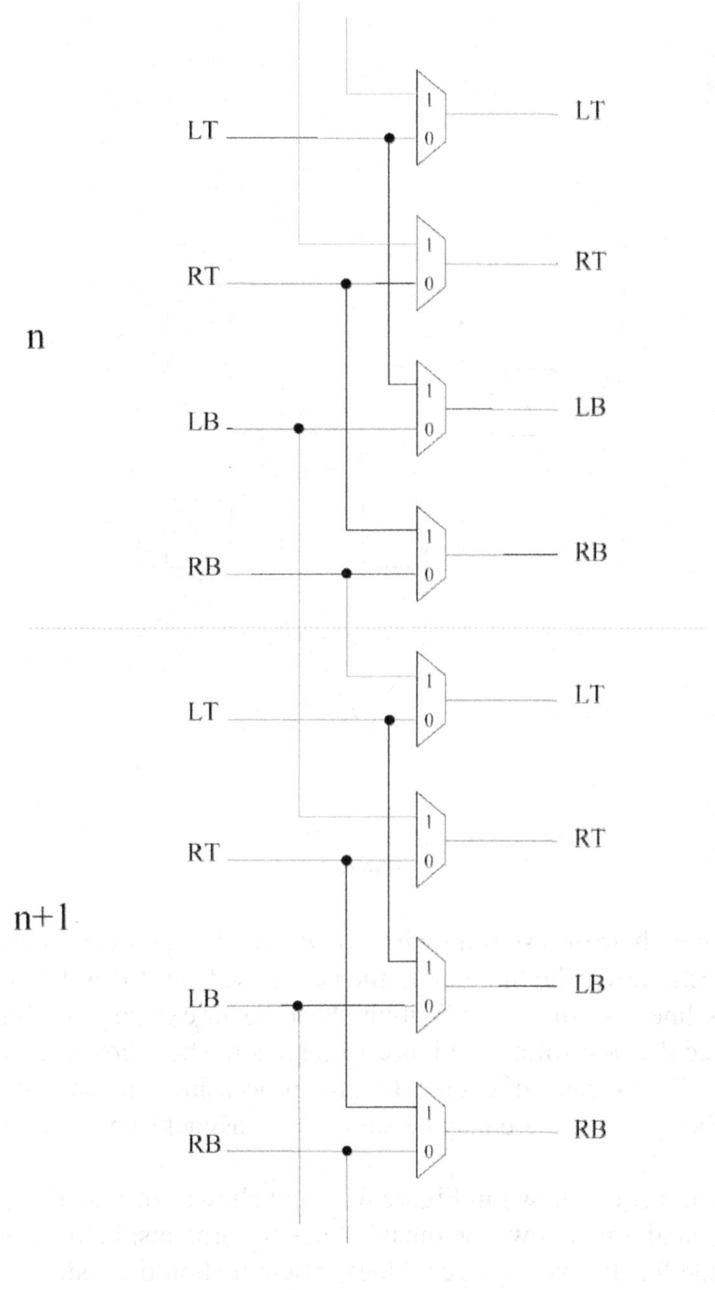

Figure 12

The routing is the same from row to row, and is shown here for rows {*n*, *n+1*}. The cyan traces connect rows {*n* – *1*, *n*}; the green traces link rows {*n*, *n+1*}; and the red connects {*n+1*, *n+2*}. All of this is needed for corner writes. You see that {LB, RB} of row *n* are being routed to {RT, LT} of row *n* + *1*. Now, the colored lines are handled *differentially* as in Figure 6, in the tap box for LB and RT (and the mirror image for link line {*m+1*} would have a tap box for RB and LT). The black and white traces, e.g. the outputs, are handled *serially*, as in Figure 6 on the right-hand side. Figure 6 shows LB to RT, but the mirror circuit would have RB to LT.

For a more intuitive picture, the following diagram shows corner locations from stagger row to stagger row, going from top to bottom. Refresh steps are indicated with numbers (see refresh steps in Chapter 17).

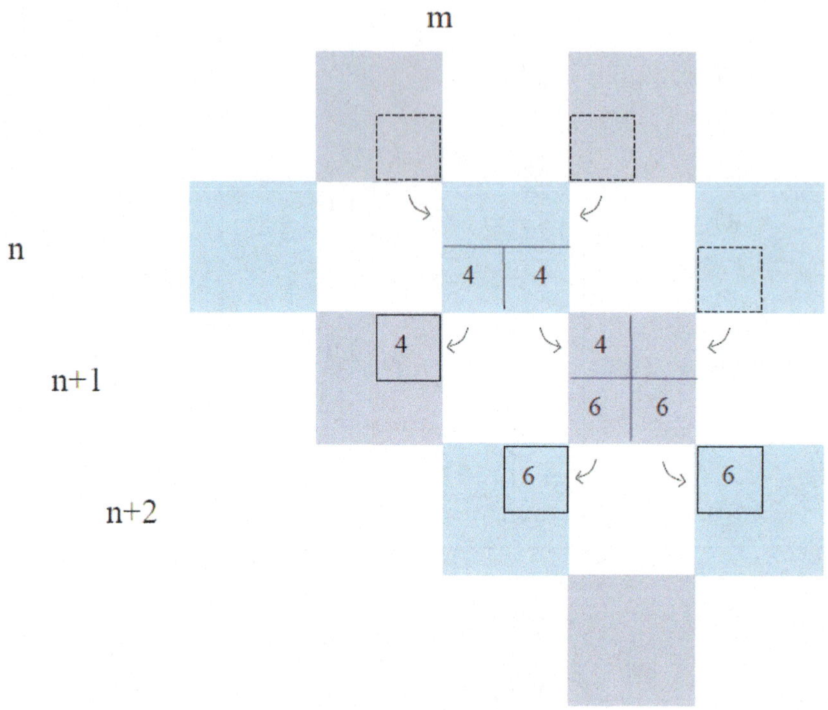

Figure 13

The K-reroutes are shown with arrows starting with row *n*, which is an even stagger. The routing logic sends the corner data directly down the link line to the next row, from LB to RT on link line *m*, and from RB to LT on the next link line over (*m + 1*). "Writing ahead" over existing link lines instead of adding auxiliary lines was deemed the best solution. Figure 13 begins to show how each center is actually served by two adjacent link lines. This means, of course, that the processing units at that center operate independently of each other because there may be some slight asynchrony from link line to link line.

The station (e.g. the switch feeder) shown in Figure 6 is next shown for *n* and *n+1* for column *m* as shown in Figure 13. Notice that on alternate rows the outside lines become inside lines, and vice versa. That is, RB is positioned on the outside for the even stagger (blue), where it should be, since the link lines are only addressing the left-hand side: LT and LB are on the inside. But one row down, for the odd stagger (purple), RB is positioned on the *inside*, closest to the odd center, and LT is on the outside, again as is appropriate. The ACK/NACK line runs through the middle of the five-line link.

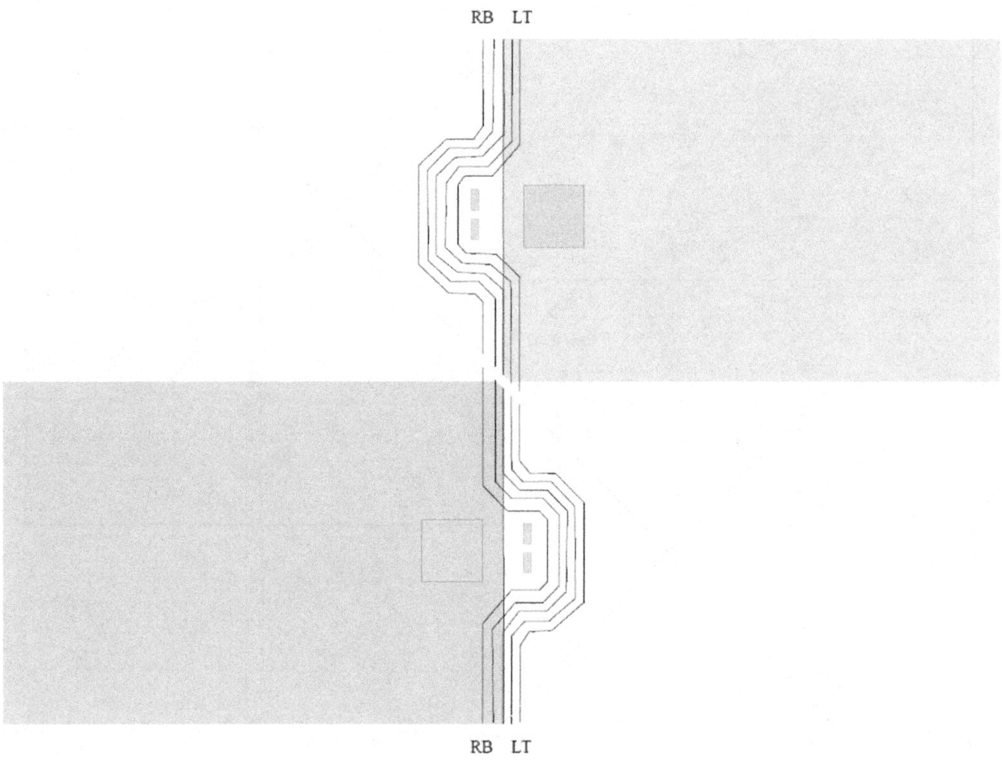

Figure 14

Because each link line serves only half a base station, the base-station circuitry must be bisymmetric, in that half the base station is served by link line *m*, and the other half by link line *m+1* and so is mirrored.

Next, the view is broadened to show 8 centers. Notice that centers only fill half the base plane. The empty squares will be made good use of however for FIFO RAM. Notice also that transparency is not a constraint at the base plane: it's understood that the cube rests on the base plane and there's no need for transparency on that one side of the cube.

Figure 15

The link lines are shown in gold, and the even and odd staggers are shown in blue and purple respectively (background color). The full base plane will consist of 8 of these in each direction, e.g. 8 x 8 x 8 = 512 centers (but centers along borders will be half disabled). Some attempt has been made to represent the scale of the circuitry, though the exact layout will of course not be determined until the switches are entered into a VLSI CAD tool and simulations are run.

Each color plane operates synchronously to a refresh clock that is in the microseconds range, as opposed to the much quicker serial clock which is in the nanosecond range. Since each link line has its own clock recovery and there might be some asynchrony, the cube is a Globally Asynchronous Locally Synchronous (GALS) system. GALS designs partition a system into multiple independently clocked domains. Reehal [6] points out that asynchrony becomes more and more unavoidable as clock speeds and chip densities increase.

A "clock network" is an optimal clock-line layout. The average power consumed by the clock network, for any one color plane, can be as high as 40 percent of the total power consumption. The clock signal for a backplane should proceed from the *geographic center*, to reduce clock skew. Typically clock, power, and ground lines are the longest "on chip" wires (in the millimeter range for typical chips) and have the longest delays. Reducing wiring capacitance reduces both wire delay and power consumption. Increasing spacing between adjacent wires reduces "coupling capacitance" which can add noise to the signal lines.

Even though a level is clocked synchronously, that doesn't mean that each column along a row will be refreshed on the same leading edge of the clock signal. The processing units operate independently, so updates may happen on the next or succeeding clock cycles. Each switch is intrinsically inactive until the refresh cycle for the column for the row begins. Still, the next row, though intrinsically inactive, does accept corner data on two of its input ports, and buffers this data in parallel (from each port), making the switch temporarily active. Still, the next row doesn't become the primary row until the four-step refresh cycle is complete for the previous row.

Switches at row $n+1$ are activated by their neighbors at row n via a "cascade enable" line. This logic is indicated by the "cascade in" and "cascade out" lines in Figure 6. The corner writes also figure into this enabling logic. The base-plane cascade shouldn't be confused with the vertical-conduit cascade, e.g. the TAXIchip logic presented in Chapter 6. In that case the serial flip flops on each color-plane backplane gate in 256 bits (64 bits times 4 processing units), and at the same time a counter is employed. When all the bits have been loaded, the next color plane up is enabled (say, the "G" plane above the "R" plane) by a cascade signal.

The voxel sequence is determined in advance by the transmitter logic in the cube holder. External memory is arranged such that any 8 x 8 window can be quickly accessed and retrieved. The transmitter logic adds the packet headers and so forth.

18.3. FIFOs

For each processing unit, each refresh step involves 192 bits, which is the full {color, *sous-bande*} information for one quad. There are 192 bits x 32 levels per level group, or 6144 bits to be stored. This times 4 for all level groups, times 4 again for the other processing units, adds up to 98,304 bits, or 12 KBytes per switch. It's not a lot of memory and a custom memory design may be effective. (The actual count is a little more due to the header information that must be stored, see Figure 9 above and Chapter 19.)

The design of the FIFOs must be simple. Since a FIFO is based on an underlying dual-port RAM, a FIFO is rather expensive both gate-wise and footprint-wise. But there's no way to clock in data at a higher speed (188.7 MHz) and clock it out at a lower speed (47.18 MHz) without buffering the data. The challenge will be to minimize the high-speed control logic, and to give priority to the essential RAM cells themselves. (See the next section for footprint ideas.)

The design in Figures 7 and 9 makes use of a level counter to route the incoming data to the appropriate level FIFOs (Figure 10). Since a full packet must be received and accepted before adding a "start pointer" to the start_ptr table, the counter is aligned to a complete packet of 64 / 8 bits = 8 bytes. The count of 8 bytes corresponds to a complete packet of "quad" data stored in the FIFOs. In other words, 8 bytes at a time are written to the various FIFOs at the 4-phase clock rate of 188 MHz. Each refresh step requires 3 packets, e.g. the FIFO addressing will not change until 3 64-bit packets have gone through. At this point, the next level-group is selected.

All four level-groups (of 32 levels each) are sent up the vertical conduit(s) in parallel, but not until the current refresh step for all groups is loaded into the base-plane FIFO memory. The time taken for writing one level group into a FIFO is the fast clock (188.75 MHz) times 192, or roughly 5.3 ns x 192 bits = 1.017 microseconds. This is the delay between each successive transfer from, for example, the LT stream, to each of its four level groups. This delay is perfectly acceptable since once everything gets going, the voxel "on" time will always be consistent for a given level, between refresh steps. At the tail end of the FIFO read cycle, a *request for send* can be sent to the optical port logic, and, with some overlap with existing data being read from the FIFO, the new burst of data can be written in.

A level switch is triggered by the data header shown in Figure 8, which is picked up as part of the data flow and triggers level logic both in the base plane switch and in a master controller at the optical port landing pad for the link line.

Finally, a note about being able to remove the cube from its holder and still have the image persist. In the future, this may be possible. The solution can be approached in two ways. First, think of the cube holder as an extension of the cube, and then, ultimately, with advances in miniaturization, imagine the entire cube holder being incorporated into the cube itself. The second approach is less drastic: simply increase the FIFO memory in the base plane from 6.144 MByte to 24.576 MByte. Then, instead of throwing away the "start pointers" retrieved from the start_ptr table in the back end of each switch, simply loop over them, since the FIFOs will contain all the data for all refresh cycles for all levels.

18.4. Footprints

Going by Figure 15 there are as many "voids" in the base plane layout as "centers". The base plane is an interesting problem in terms of footprint because of its NoC character. Each center base station is about 1.5 mm squared. While the occupied center area will accommodate the RAM *control* circuitry, the RAM cells themselves must go elsewhere. The FIFO memory is the last bit of real estate that needs to be resolved, but notice in Figure 15 that half the base plane is virtually empty. Below, the same graphic is extended to show how 4 FIFO RAM arrays per processing-unit quadrant per center neatly fit into those empty spaces.

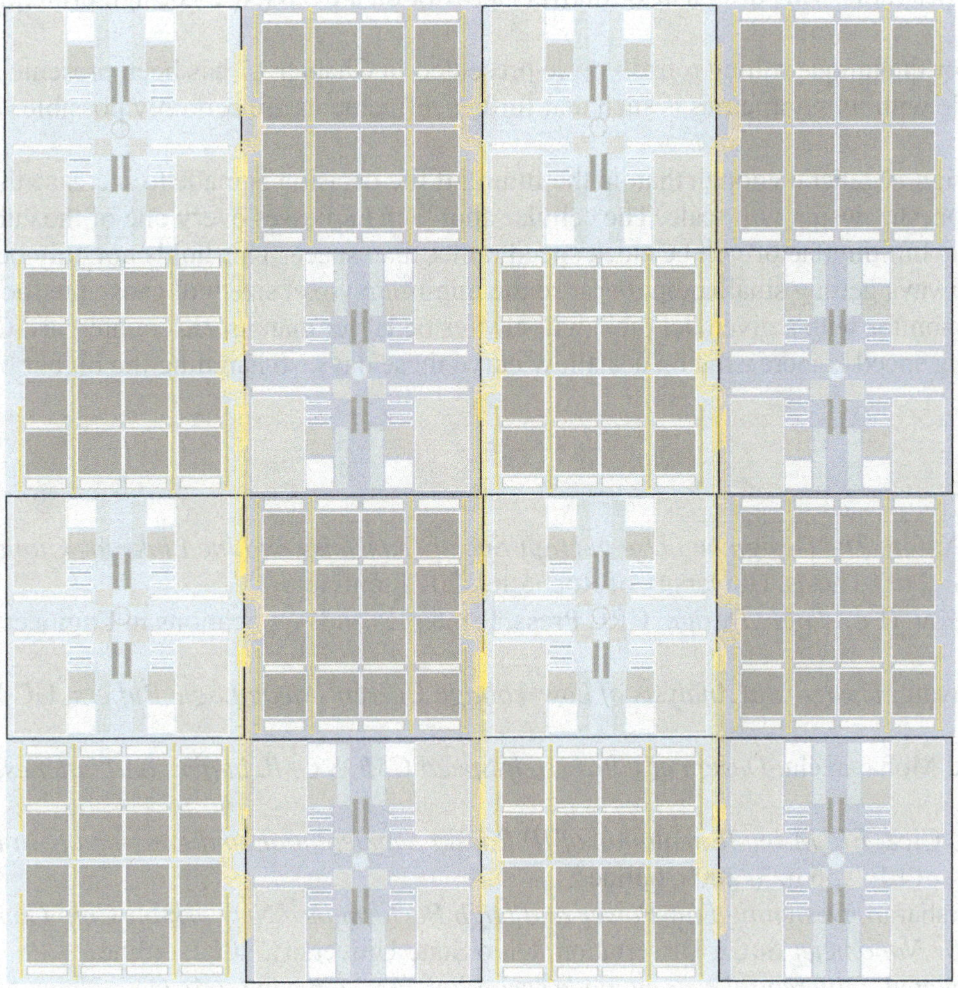

Figure 16

Notice that the RAM adjacencies are in the vertical direction, so that the memory lines don't interfere with the link lines (in gold), also running vertically. Each "memory center" (the 16 FIFOs) is split between two neighboring switch centers in the vertical direction, above and below. The entirety of Figure 16 measures roughly 6.25 mm to a side: the circuitry is *very small.* The entire base plane consists of 64 copies of Figure 16 put together into an 8 x 8 grid. For clarity only 3 link lines are shown running vertically (gold). The left/ right boundaries when fitted together would add a fourth link line.

18.5. Conclusion

This chapter has looked at the base-plane design for the 3D display cube in more detail. Because the base plane is the same size as the actual display-area footprint, the total base plane area is limited to about 2 inches or 50 mm squared. But even with this limited space, the full design as envisioned can be accomodated.

FIFO memory in the base plane acts as a buffer for the high-speed data arriving via link lines connected to the decoder circuits at each optical port. The FIFO requirements combined with the need to manufacture the base plane as a single piece, suggests that a custom FIFO design that's tightly integrated with the base station logic is desirable. This design also follows Network on a Chip (NoC) design principles.

Control logic for the unique rolling refresh cycle presented in Chapter 17 has been presented. The nature of the refresh cycle with its symmetries is such that further refinements are certainly possible.

Finally, the design is general enough that, in the future, if the decision is made to increase the display cube's resolution, the design will scale. The cellular approach harnesses every one of the cube's three dimensions. The limiting factor will be clock speed rather than space. This holds not only because fabrication is always getting smaller, but because the minimum voxel size will cause the footprint to widen, much as a 4K monitor with a given dot pitch will always be larger than its 2K counterpart. On the other hand, processing speed is increasing on a shallow curve these days, so handling the bigger bandwidth will be a challenge.

References

[1] Carpenter, Aaron. *The Design and Use of High-Speed Transmission Line Links for Global On-Chip Communication*, PhD Thesis, University of Rochester, 2012, online
[2] Chen, Wai-Kai. (Ed.) *Logic Design*, CRC Press, Principles and Applications in Engineering Series, 2003
[3] Heydari, Payam. *Design and Analysis of Low-Voltage Current-Mode Logic Buffers*, UC Irvine, 2003, online
[4] Heydari and Mohavavelu. *Design of Ultra High-Speed CMOS CML buffers and Latches*, UC Irvine, 2003?, online
[5] Carevic and Cica. *FPGA Implementation of IP Packet Segmentation and Reassembly in Internet Router*, Serbian Journal of EE, v. 6 n. 3, 2009, online
[6] Reehal, Gursharan. *Designing Low Power and High Performance Network-on-Chip Communication Architectures for Nanometer SoCs*, Dissertation, Ohio State University, 2012, online
[7] Esper-Chain et al. *Differential current mode serial link for 2 Gb/s in GaAs technology*, Microelectronics Journal, v. 33, p. 1107, 2002, online
[8] Kim, Jongsun, and Bae. *A 2 – 4 GHz fast-locking frequency multiplying delay-locked loop*, IEICE Electronics Express, v. 14, n. 2, 2017, online
[9] Kim, Jongsun, and Shin, Hyungsik. *A Low-power 3.52 Gbps SerDes with a MDLL Frequency Multiplier for High-Speed On-chip Networks*, Journal of Semiconductor Technology and Science, Dec. 2018, online
[10] Nebhrajani, Vijay. *Asynchronous FIFO Architectures, Parts 1, 2, and 3*. Available online (three separate downloads from different sites).
[11] Ramesh, Kumar, and Reddy. *Asynchronous FIFO Design with Gray code Pointer for High Speed AMBA AHB Compliant Memory Controller*, IOSR-JVSP, v. 1, issue 3, 2012, online
[12] Pinkston and Duato. *Interconnection Networks, Computer Architecture: A Quantitative Approach*, 4th Edition, Appendix E, ppt online
[13] Alexandropoulos et al. *A novel 1.8 V, 1066 Mbps, DDR2, DFI-compatible, Memory Interface*, 2010 IEEE Annual Symposium on VLSI, online

Chapter 19

FIFO Design

19.0. Introduction

In this final chapter the subject of storage, specifically RAM, is addressed. Let's jump right in.

Recall that the display cube has very fast serial data clock running at 755 MHz. That clock is divided by four, to 188.75 MHz. The period drops from 1.324 ns to 5.298 ns. The data is distributed over 4 serial lines instead of 1. The data for all four processing units is sent up the vertical conduit in parallel. So far so good. But recall that the data is sent to 128 levels. The decision is made to send four parallel streams to 32 levels each. This divides the clock by four again. The clock frequency for delivery up the vertical conduits becomes 47.187 MHz, with a period of 21.19 ns. The slower clock rate avoids the use of differential logic (such as the CML in the base plane) in the color planes and conduits. The slower rate also contributes to a 25% duty cycle for the LED refresh, which would otherwise be hard to achieve.

At each base station the FIFOs (first-in first-out memory) serve the purpose of buffering the fast data by accepting the fast data on one side and gating it out on the other at the slower clock speed for delivery up the vertical conduit. The FIFO is the critical piece of the whole design.

19.1. RAM

The four-phase optical recovery creates 4 processing-unit link lines that replace a single link line and operate in parallel. Each of the four lines is a point-to-point differential link. Each link line leads to a succession of base stations, each of which serves as a high-speed switch. A traditional routing switch uses a demultiplexer (one to many) to accept incoming data and make virtual packet streams from a single input. If there are four virtual packet streams then the data is buffered to four different FIFOs. Each FIFO output is then routed to an input on four different multiplexers (many to one). The multiplexers can then select any one of the four virtual streams, in parallel, for routing to available destinations.

The cube routing is more determined and less dynamic than a general routing switch, so the multiple-choice design isn't needed. But the four signals do need to be distributed to 4 *level groups*, leading to 16 lines total. And because the output clock is four times slower than the input clock, an asynchronous FIFO design is needed. More, a FIFO is needed for each level group, making 16 FIFOs total per switch, 4 for each processing unit.

A FIFO is a dual-port RAM, and it pays to go into this a little bit since knowledge of how memory works will lend insight to the FIFO design. The basic 2D memory architecture is shown below. 2D refers to the cell layout. There are also 1D serial memories, and even 3D memories.

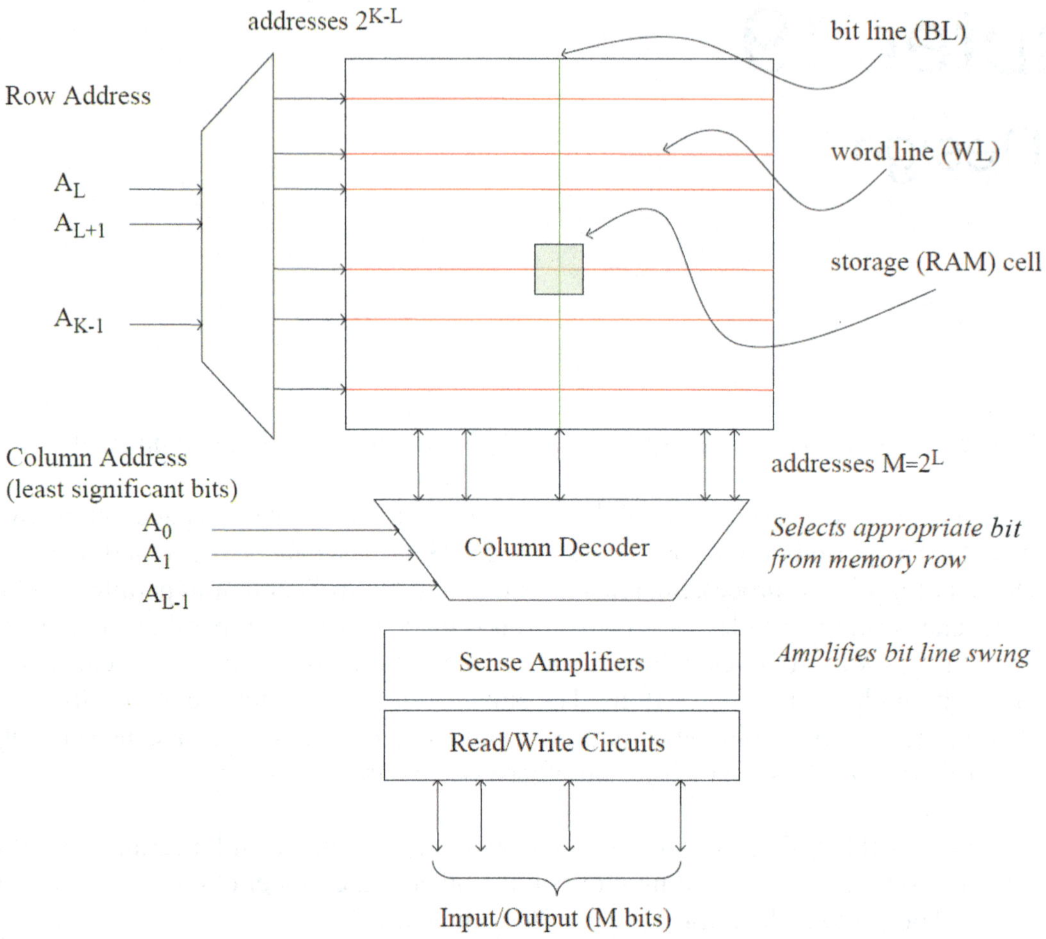

Figure 1

Memory is referenced by row and column. The main thing to notice is that only one cell is addressed. A cell in Figure 1 is one bit, but in real life, even though a row is composed of bits, there can be several columns written to or read from at once, in which case the column driver is said to be *distributed*. For the one-bit design shown above, if the RAM word size is M = 16 bits then the column address has width L = 4. Each bit would be read out or written in during a clock cycle, so that a minimum of 16 clock cycles would be needed to fish out the bits. Again, this isn't too realistic and a better approach will be given presently.

This means 16 bits reside on a row, and there are K – L rows. If K is 12, then there are 2 to the power of (12 – 4 = 8) rows, or 256 rows. These parameters are design-dependent.

Decoders are addressing units which activate a numbered line. The selection value, a row or column address in this case, is used to activate a memory cell. The decoder boxes contain logic similar to the simple 2-bit decoder shown below.

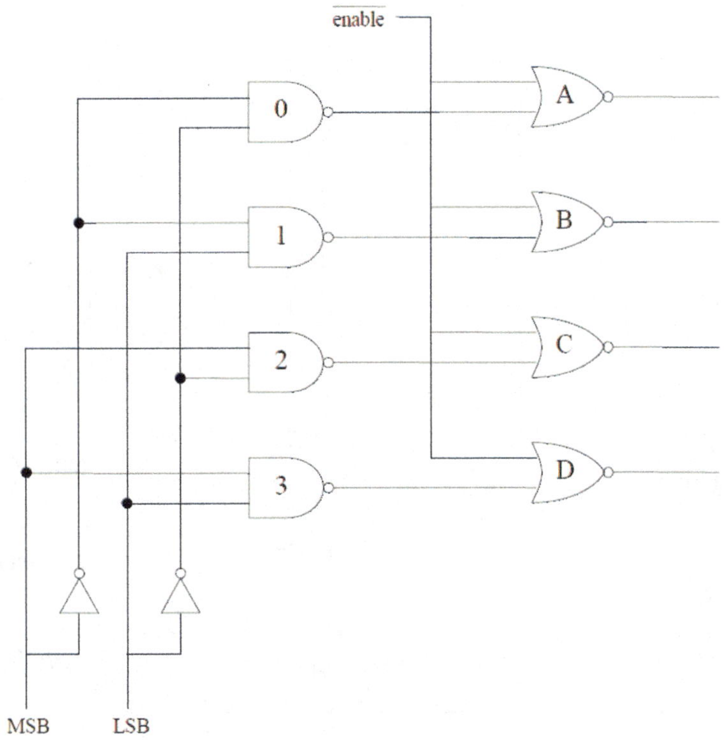

Figure 2

This is the kind of circuit that's taught in a freshman logic design course. In CMOS logic each 2-input NAND or NOR gate requires 4 transistors. The logic table below shows how the 2-bit address is combined with the enable (active low) to select (high) one of four banks {A, B, C, D}.

MSB	LSB	0	1	2	3	en	A	B	C	D
0	0	0	1	1	1	0	1	0	0	0
0	1	1	0	1	1	0	0	1	0	0
1	0	1	1	0	1	0	0	0	1	0
1	1	1	1	1	0	0	0	0	0	1
x	x	x	x	x	x	1	0	0	0	0

Table 1

The diagram below shows a dual port version of Figure 1.

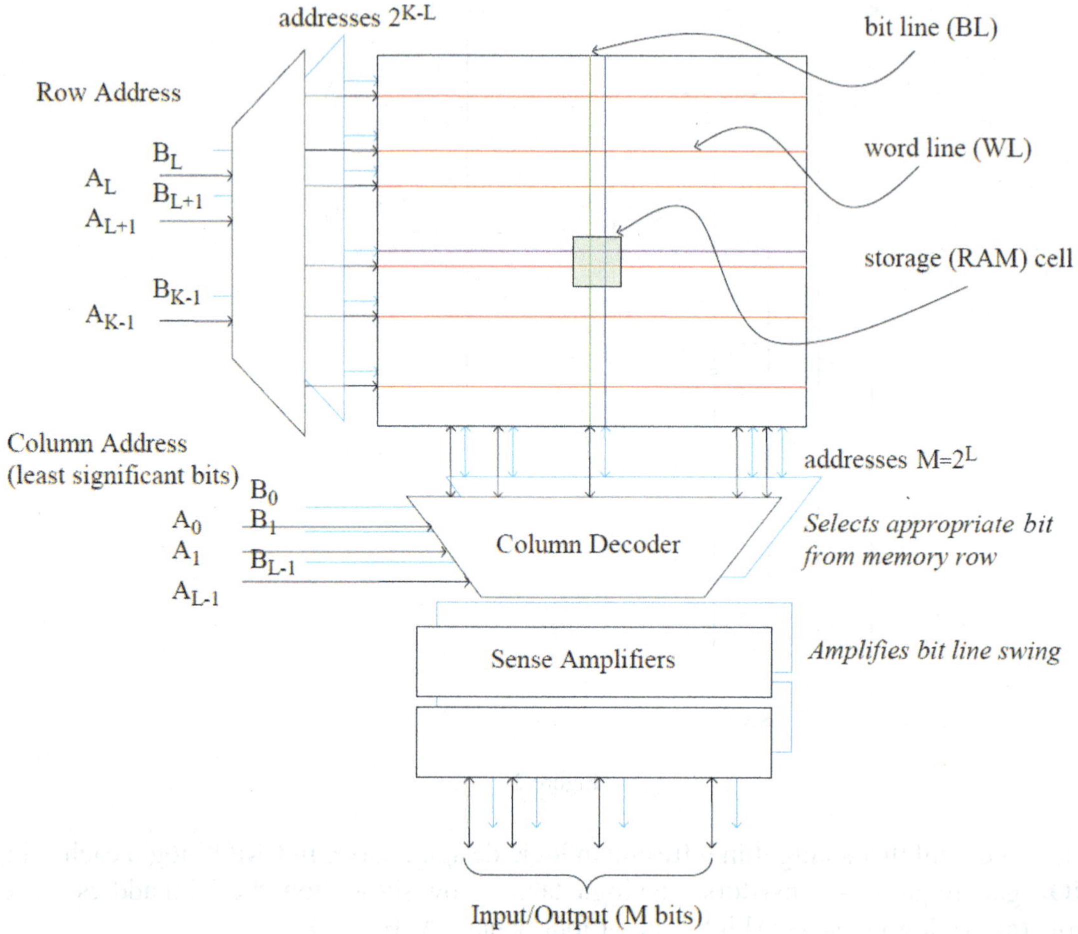

Figure 3

There are two sets of address lines for row and column, and two sets of read/write logic. But internally, there are also two sets of word lines and bit lines: they can be labeled as A and B. The contents of each set are identical, so that the contents whether read from A or B will be the same. Using the familiar language of transistors, a dual-port one-bit latch is shown below.

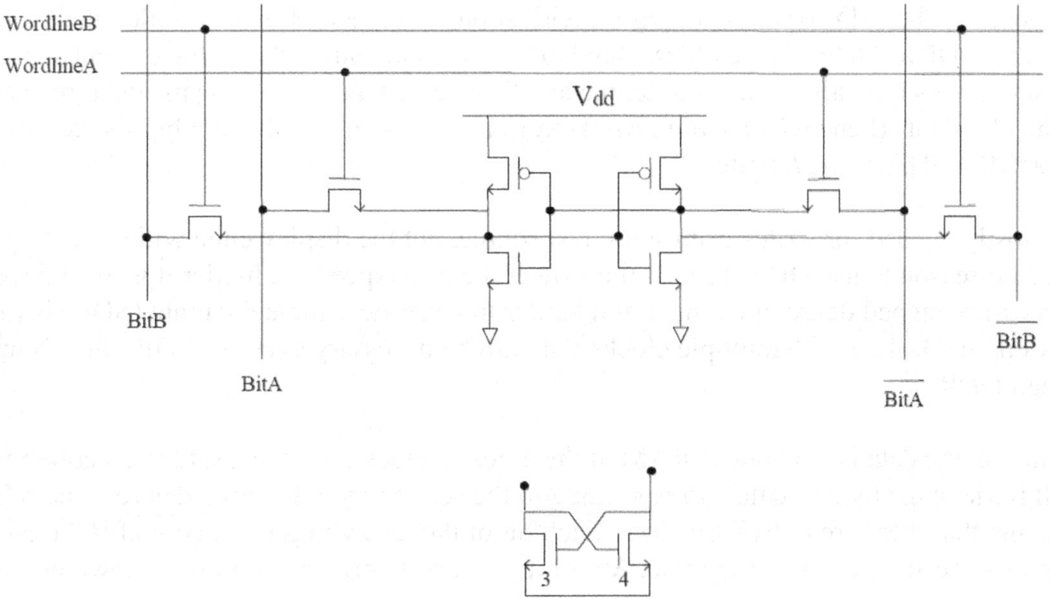

Figure 4

The logic is differential. The middle part shows the *primary storage nodes*. Notice that the lefthand side is mirrored on the righthand side, making the storage nodes two-sided or dual. The bit lines A and B act as the data lines. At the bottom a cross-coupled latch is shown for comparison, showing how the RAM cell is essentially a latch. Notice how the BitB and BitB_bar lines, along with WordlineB, essentially frame the regular A logic. If you remove all of the B lines, you end up with a standard 6T (six transistor) cell.

The manufacture of RAM chips by companies like Micron and Samsung involves layers and capacitance and redundancy, so Figure 4 is only the first part of the design; the design must be finalized with transistor widths and so on.

Figure 4 is the innermost sanctum of the memory path. There are two main types of RAM: static (SRAM) and dynamic (DRAM). A DRAM cell (omitting the refresh circuitry) is simpler than SRAM, and for thousands of cells this can result in a space savings. But SRAM, though having a slightly larger footprint, is faster than DRAM. DRAM requires refresh circuits and storage capacitors, so is more complicated.

19.2. Physical Interface (PHY)

Outside of the RAM proper the incoming data has to be prepared. This happens at the physical interface level, otherwise known as the PHY. [4] It's instructive to go through the read operation first. To start, a more sophisticated RAM design will still address bits in each row, but a column address of, say <0:0:0> will select not just one bit, but 8 bits all tied to that select line. The bits represent bit positions or *bitslices* <7> down to <0>. All 8 bits will arrive at the data bus simultaneously. For example, the array can be arranged as 8 bytes of $n = 8$ bits each, or $N = 64$ bits per row, and, say, $m = 8$ rows. This array can also be called a *core*. Given a core width of $k = 8$ bytes, the core size is $k * n * m$ or $8 * 8 * 8 = 512$ bits.

The RAM being read returns 8 bits of data on 8 different data lines. Those 8 bits come into the PHY on a read_DQ data line corresponding to the bitslice position. The PHY is normally *bidirectional*, meaning the same lines are used for writes.

One of the purposes of the PHY, besides getting individual bits to the RAM, is to double the data rate – the mnemonic for this is DDR. DDR2 is the oldest standard still extant, and DDR5 is the current standard. DDR2 and its successors basically create a 0-degree and 90-degree pair of phased signals to read and write data at twice the clock rate (hence *double data rate*). Normal operation speeds for a high-speed RAM are from 500 to 800 MHz. This is the *bit* rate.

High speed is mostly needed for writes from some fast source, but the display cube will be writing at 188 MHz, so there's no reason to use DDR. In fact, the slower the clockspeed the harder it is to achieve DDR because the necessary tapped delays are longer and harder to generate. Single data rate (SDR) is fine, at least for our example. Dealing with multiple clocks will still be necessary even for SDR, but it's much less complicated than DDR.

In the read example, the data is read out of RAM at the external clock rate. For example, a column address of <0:0:1> will retrieve the byte of data corresponding to the second byte distributed across the selected row. Now, imagine that there are 8 PHY bitslices. Each bit of the data will go to its own PHY bitslice. Each data slice consists of read logic and write logic. An 8-bit physical interface diagram is shown below.

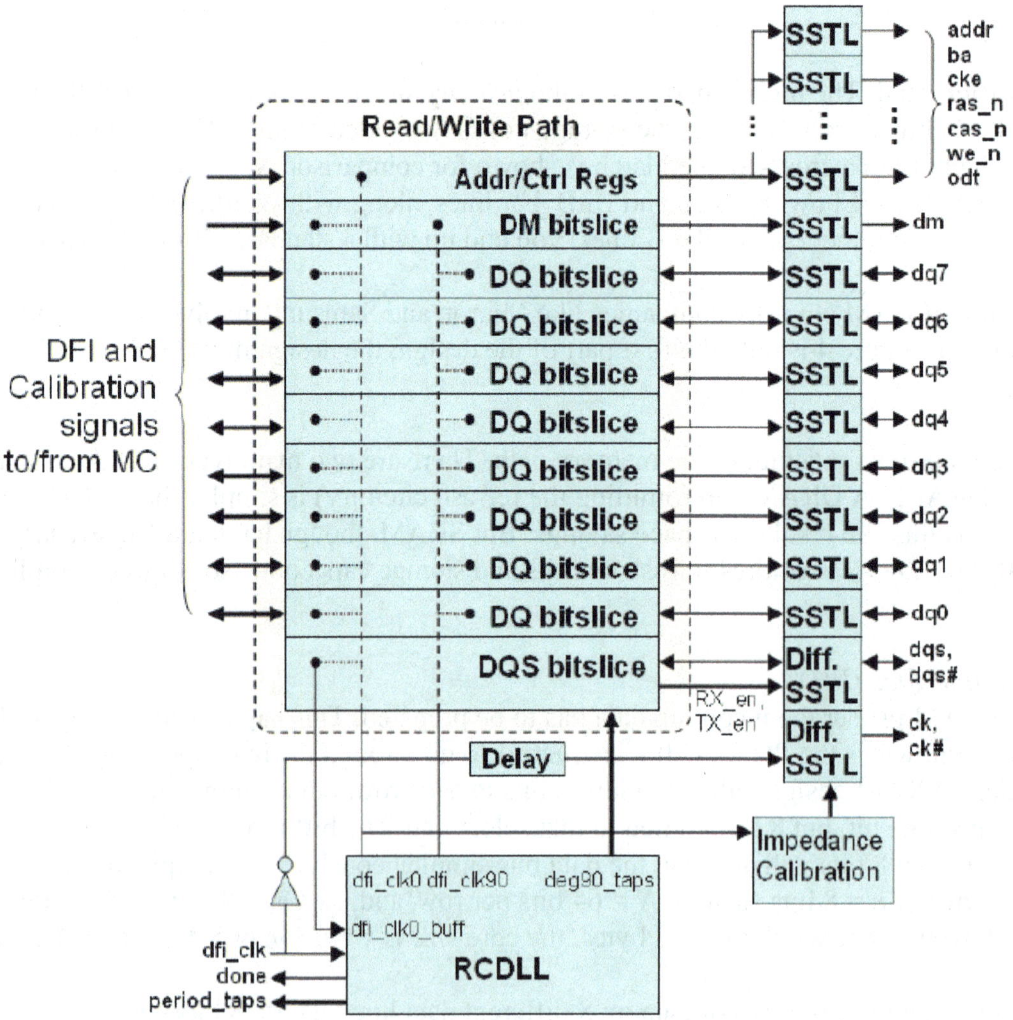

Figure 5. (Courtesy [4])

Given 8 bitslices corresponding to an 8-bit memory word, for a write operation the data arrives as a byte, and is routed through the PHY, each bit going to its own bitslice. Each bitslice transmits the bit to a write_DQ line that is directly connected to the RAM bitline. The idea is to match the RAM word width (being 1 byte or 8 bits in our example) to the number of slices so that the data lines have *point-to-point connectivity*. All 8 bit slices are gated into RAM for the current column address (say <0:0:0>) and row address.

A DDR PHY generally consists of command and data slices. The commands go through the PHY command slice to memory and the read and write data goes through the data slices. MC in Figure 5 stands for Memory Controller, which is the high-level logic to the *left* of the PHY; while the RAM can be visualized as being to the *right* of the PHY. The PHY translates incoming read- and write-enable signals from the MC into IO buffer enables and strobes (called *write DQS* and *read DQS*) which gate the data.

DQ refers to the DQ flipflop logic used by the PHY. The tricky part of the design is the RCDLL at the bottom of Figure 5, or Register Controlled Delay Locked Loop. A delay locked loop (DLL) is a circuit similar to a phase locked loop (PLL). The main difference is that the internal voltage-controlled oscillator is replaced by a delay line. The DLL is needed for converting single data rate to double data rate. But working only with SDR, there's still some locking that needs to be done when the data arrives in bursts. And here's where things get tricky.

Let's say that, from the outside world, 8 bytes arrive in sequence. Now each data line looks like a series of bits all coming in sequence, one after the other. Even at 188 MHz, special clocking is needed to extract the data from each serial stream. If there are 8 bitslices in the PHY, there will be 8 serial data streams. From a bird's eye view, this data just looks like a lot of activity on the external data bus. But it's actually a stream of bytes on the data bus all coming in one after the other.

In our example, there are 8 bytes per row, so each row can handle a stream of 8 bytes, as just described. The PHY has to extract each bit from the incoming stream, and, because all 8 bits for byte 0, for example, must arrive and be gated by the same column address for all 8 bitslices, the timing is such that the transfer of bits must be synchronous. Each bit of byte 0 must be extracted from each of the 8 data lines using the same clock edge.

Let's address a detail now, being the *interleave* of the data bits across a RAM row. The decoder has a 3-bit address input, and will activate one of eight output lines, similar to Figure 2. Call each output line, or column driver, *y*. The easiest silicon layout has *y0* going straight across the array. This is shown conceptually below, for 3 columns corresponding to 3 4-bit bytes in each row.

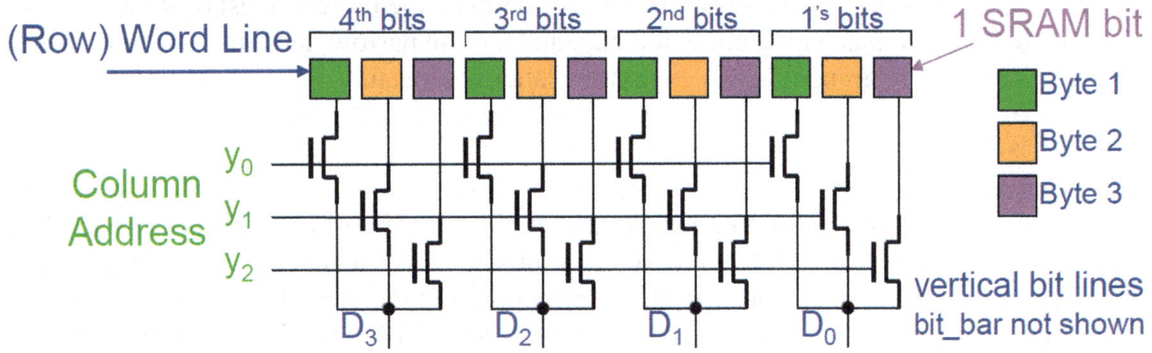

Figure 6 (Courtesy [6])

For the PHY example of 8 8-bit bytes, bit <7> is the first position; bit <6> is the ninth position; bit <5> is the seventeenth position; down to bit <0> at the fifty-seventh position. This is called an *interleave*. The line driver runs horizontally and the data lines run vertically. This allows the second line driver for column <0:0:1> to access positions 2, 10, 18, … , 58. These are the bits of byte 1. The column address is also fed to a read MUX, which selects, for each data output, one of those 8 contiguous positions. Therefore, column 0 references byte 0, column 1 byte 1, and so on, up to column 7 byte 7.

This is how the 8 bytes arriving in a stream are stored. They are stored right across the current row based on the current column select. The column address will cycle from 0 to 7 for each bit position in the stream, until all 8 bytes are written. It's possible for 8 bytes at a time to be read out of RAM as well, but the normal operation is simply to read 1 byte with a single column address.

SSTL in Figure 5 stands for *stub series terminated logic*. This is an electrical standard for driving the transmission lines (read, write, and command lines) used by the RAM. SSTL is a high-speed specification that can handle the DDR signals transmitted over the data bus. Typically the RAM has no output register for the read data, and instead the data is strobed onto the SSTL bus, each line of which directly connects to the appropriate PHY bitslice. Note that the timing is in the nanoseconds – this bus is very fast.

A PHY interface can control multiple ranks of RAM chips with multiple chip select signals. This assumes that each RAM core is on a separate chip. However, forget about multiple chips and instead think about multiple RAM cores, all on the same die. The display cube will have all modules for a switch on the same silicon – indeed all of the switches and their memory are on a single die, as shown in Chapter 18.

So the complexity of mounting chips with different timings due to differences in the processing of each DRAM die are not relevant for the display cube, which simplifies the PHY. The idea of a chip select can be retained, but the naming needs to be changed from *chip* select to *core* select. In our example there's a 512-bit core. To add more memory more cores can be added. This idea goes back 40 years – really not much has changed in the RAM world besides speed and density.

Let's say we add 8 cores in a pattern of 4 rows of 2 cores each. The original row address lines {2, 1, 0} will be routed to every one of those 8 cores, as with the single core. But now we add new row address lines {4, 3} which will select each row of 2 cores. Let's say we want to preserve the word size of 8 bits. Then the original *column* address lines {2, 1, 0} are routed to all 8 cores and an additional column address line {3} will select between the left core column or the right.

On the other hand, let's say we want to increase the number of PHY bitslices from 8 to 16. In that case, the additional column address line is unnecessary, because the second column of cores will be connected to a whole new set of bitslices, from <8> to <15>. The wider the PHY the more real estate it takes up. There's a tradeoff between a wide data bus with parallel loading, and a more narrow data bus with more data arriving serially on fewer lines. We're going to try to exploit the latter approach.

19.3. PHY Redesign

The PHY in [4] occupies a footprint of 1 mm squared using 90 nm technology. This is much too large to export to the cube switch landscape. Recall the MDLL footprint of 200 x 800 mm (Chapter 7). This is okay repeated 32 times at the landing pad, once for each optical port, but the cube design simply cannot accommodate either the power profile nor the real estate for the MDLL nor the PHY in the many base plane switches.

The most important part of the PHY is the read/write path. The PHY is very good at processing several bytes arriving in a stream by picking off individual bits for each bitslice dataline. The thinking is that, say, 8 bytes can be read directly out of RAM and into a shift register, one for each bitslice, and then clocked out of the register at a slower speed for transmission up the vertical conduit. In other words, the PHY can also serve as the output processor.

Since each color plane takes 16 bits x 4 voxels = 64 bits, a serial burst of 64 bits makes the most sense. A byte can be any width. If we made it 3 bits for 3 color planes, we'd run into trouble because all 3 colors would have to be presented at once. Instead, make the input 1 bit only, being the serial data line with minimal delays and stripping of header packets. The three colors are then routed to separate rows in the core. Let's set it up as 2 cores across, each with 32 x 1 bits = 32 bits per row. That way all 64 bits can be written to or read from RAM.

The 64-bit burst will require cycling through a 6-bit column address space. The low-order column address {4, 3, 2, 1, 0} is routed to all cores and selects one of 32 bit positions. The high-order column address {5} acts as the core_select signal, and selects the left core or the right.

How many rows? Rows will be ordered by *color* first, and then by *levels* in the cube. First, we need to consider the 4 processing units. But even before that, we need to consider the 4 level groups, which are read out in parallel. Let's increase the core array size to 2 x 4, with the 4-dimension corresponding to the 4 level groups.

Before going on, recall that there are 4 processing units and the processing unit data arrives in parallel from the 4-phase MDLL at each optical port. So our nominal array of 2 x 4 cores is duplicated 4 times, one array for each processing unit. Each p.u. array operates independently.

Back to level groups. Each 2 x 4 array of cores will have a *unidirectional* read bus. Each row of 2 cores on the read side consists of a single data line which handles a read burst of 64 bits, one for each color. Since there are 4 level groups, there are four data lines on the read side. Each data line is unidirectional and routed into a 64-bit shift register, one color at a time. The 4 shift registers operate in parallel and represent the level groups.

On the write side, the 64-bit serial data arriving at high speed is kept in serial form and synced to the column address, which changes for each bit in the bit stream. The 4 core rows for the level groups are written to in sequence as the data for each level group arrives. The order is {red, s.b} x 4 voxels (quad), which is 64 bits, followed by green and blue, followed by the same ordering for the next 3 level groups, before repeating for the next level in each level group. Each core will have 3 colors x 32 levels = 96 rows. Full row addressing amounts to: low-order {6, 5, 4, 3, 2, 1, 0} for 96 rows (32 unused) routed to all cores; and high-order {8, 7} for the level group, e.g. the 1 of 4 select for one of the 2 x 4 rows. The write order is all level groups for a level, for all levels. This means the high-order bits change the fastest.

The outermost loop of the write order is the refresh steps, being one of {Y, C, X, K} not necessarily in that order. Each refresh step cycles through the levels as just explained. If the FIFO isn't *mostly empty*, the write burst with the new refresh step will have to wait until the base-plane switch sends a *request for send* over the link line status line. Then the new entries will arrive as a new sequence of levels for the next refresh step.

Does it all add up? Each core width is ($k = 32$) * ($n = 1$) = 32 bits and each core row-size $m = 96$. We have 32 bits x 96 rows = 3072 bits per core. There are 2 cores for each level group, giving 6144 bits. This times 4 for all level groups, times 4 again for the other processing units, adds up to 98,304 bits. This memory is contained in 32 cores, 8 cores per processing unit. Note that bitwise the core's footprint will not be 2 x 4, but more like 32 bits x 96 rows on the left, control circuitry in the middle, and 32 bits x 96 rows on the right, with the 4 cores in each 2 x 4 core array stacked vertically. The 4 read-data lines would travel in the same direction as the serial links, vertically, to the vertical-conduit shift-register load port. The layout for the 4 processing units would be 2 core arrays on top, and 2 core arrays below.

Other details. As shown in Chapter 18 Figure 9, each 64-bit packet is associated with a header which contains the level number for some level group. Level changes have to be recorded for the cascade logic that sends cascade signals up the conduit. It's safe to include the header info as a 5-bit code (for 32 levels). There's also a stop bit that indicates the transition from color to color. This will be needed for the vertical cascade logic that advances from color plane to color plane. So 6 bits need to be added. The level group should also be included, making the total word width 72, with 36 bits per core. The 8-bit header *leads* and the 64-bit payload *trails* in the serial stream, so the header would be immediately captured from the read bus in its own register for the conduit logic.

The conduit logic will also require 2 bits for the refresh step. The refresh step tells each processing unit exactly which voxels to address. However, it may not be necessary to store the refresh step in FIFO memory, but simply generate it in the control circuitry at the appropriate time. This is possible because the refresh step is aligned with ACK/NACK/RTS, e.g. each packet burst is aligned on a refresh step. Compared with the other parameters, the refresh step changes the least. To give the LEDs time to persist, the actual write order is {even staggers, odd staggers}: {step 1, col 0}, {step 1, col 2}, … ,{step 1, col 30}; {step 2, col 0}, {step 2, col 2}, … ,{step 2, col 30}, … ,{step 4, col 30}; {step 1, col 1}, {step 1, col 3}, … ,{step 1, col 31}, {step 2, col 1}, … ,{step 4, col 31}. Repeat. (See Chapter 17 for timing details.)

19.4. Discussion

Some design considerations for the RAM as a FIFO are now summarized. The theory of high-speed FIFOs is well covered in the three papers by Vijay Nebhrajani [1]. Additional resources [2, 3, 5] are also helpful. Nebhrajani uses the VHDL specification language to proof the logic needed to run the FIFO as a field programmable gate array (FPGA). The VHDL code is included in the Appendix, along with a VHDL spec for a dual port RAM. Both specs are simplistic but give a good idea of the kind of logic required for the FIFO and its RAM core. Suffice to say that having a FIFO with one frequency on the input clock, and a second frequency on the output clock, is a well understood problem with well defined solutions.

The main concern with dual port RAM is *metastable states*. If a signal is sampled during a state transition, is the value 0 or 1? The problem is that the value of a counter is being sampled with a clock that's asynchronous to the counter clock. If every single bit goes metastable, the FIFO won't work. But something called a *gray code* can be used.

19.5. Gray Code

In an asynchronous and in some free-running synchronous designs a component called a *status block* performs operations on two different pointers – read and write. Each pointer is synchronized to a different clock, either a read clock or a write clock. The two clocks make the ports of dual-port RAM asynchronous.

The basic rule is that the *write pointer* points to the location that will be written next, and the *read pointer* points to the location that will be read next. A write post-increments the write pointer, and a read post-increments the read pointer. A FIFO can be FULL or EMPTY depending on the values of these two pointers.

In a synchronous design you sample the read pointer with the *read clock*. You sample the write pointer with the *write clock*. In an asynchronous design, a *gray code* is used for the count. When using a gray code only one bit changes between one count value and the next. This guarantees that "counting up" will only affect one bit at a time. This means that if a counter's actual value changed from $n - 1$ to n, the value read will be $n - 1$ or n, but no other value. This makes all the difference when computing FULL and EMPTY states for the FIFO.

The value for a read will always be retrieved from position n or $n - 1$, and choosing $n - 1$ makes the count *pessimistic*. The read or write is safeguarded because the gray code makes it impossible to retrieve the wrong data or to write to a full buffer.

WRITE: the gray-coded *read pointer* is synchronized to the write clock. At worst, the read pointer is $n - 1$ (stale). That is, the read pointer has gone metastable on $n - 1$. The write thinks there are *more* entries in the FIFO than there are. The write will not write. A *full signal* will block until the correct read pointer value comes through.

READ: the gray-coded *write pointer* is synchronized to the read clock. At worst, the write pointer is $n - 1$ (stale). That is, the write pointer has gone metastable on $n - 1$. The read thinks there are *fewer* entries in the FIFO than there are. The read will not read. An *empty signal* will block until the correct write pointer value comes through.

Conversion to gray code only happens inside of the *status block* circuit for both read and write. Otherwise, binary counts (addresses) go directly to the dual-port RAM decoders: the inputs are binary, not gray code. The gray code is incremented only inside of the status block which is timed to the write and read clocks, and helps determine EMPTY and FULL flags.

Example. Let's take a simple 4-bit address width. If $n = 4$ bits (the FIFO depth) we have 16 values. Using the "+" symbol for a logical XOR (exclusive OR), and ">>" for a shift right, the gray code generation from the binary is:

gray = bin + (bin >> 1)

Now, XOR is:

$0 + 0 = 0$
$0 + 1 = 1$
$1 + 0 = 1$
$1 + 1 = 0$

So:

$0 + 0 = 0000$
$1 + 0 = 0001$
$2 + 1 = 0011$
$3 + 1 = 0010$

4 + 2 = 0110
5 + 2 = 0111
6 + 3 = 0101
7 + 3 = 0100
8 + 4 = 1100
9 + 4 = 1101
10 + 5 = 1111
   10: 1010
   5: 0101
11 + 5 = 1110
12 + 6 = 1010
   12: 1100
   6: 0110
13 + 6 = 1011
14 + 7 = 1001
15 + 7 = 1000

and then repeat. Notice how only one bit changes from *n* to *n + 1*. To convert gray to binary, and using 15 and 11 as examples:

```
mask = num >> 1   # num is the gray code to start, but is converted to binary
while mask != 0
   num = num + mask
   mask = mask >> 1
end
```

mask = 0100	mask = 0111
1000 + 0100 = 1100	1110 + 0111 = 1001
mask = 10	mask = 11
1100 + 0010 = 1110	1001 + 0011 = 1010
mask = 1	mask = 1
1110 + 0001 = 1111	1010 + 0001 = 1011
1000 => 1111 (15)	1110 => 1011 (11)

Table 2

Remember that FIFO stands for "first in first out" so the first data value written *in* is the first one read out. Now, let's say that the read pointer is at 0, meaning nothing has been read out yet. Let's say the write pointer is also 0. This is the start state. The FIFO is EMPTY.

The pointers point to the location to be written to or read from *next*. The pointer math is always done on the post-write incremented write pointer, and the post-read incremented read pointer. Let's say we've just written to 14 and the write pointer is incremented to 15, and the read pointer is still 0. Note that comparisons are in binary.

read pointer: 0000 (0000 gray)
write pointer: 1111 (1000 gray)

The pointers aren't equal, so the FIFO isn't full. Now, write to 15 and increment the write pointer. The write pointer rolls over to 0 because of the limited bit-precision.

read pointer: 0000   (0000 gray)
write pointer: 0000   (0000 gray)

The pointers are equal, so the buffer is FULL. The rule is: if the cause of pointer equality is a write, the buffer is FULL. If the cause is a read, the buffer is EMPTY. In this case, any write about to come through is blocked by the FULL status. Let's say the read pointer is really at 0001 and it will come through on the next write clock because the read state is no longer metastable.

read pointer: 0001   (0001 gray)
write pointer: 0000   (0000 gray)

The values are no longer equal, so the buffer can be written to. The write pointer is incremented to 1, and so on.

The gray code will advance to 1000 for 15, and if the pointer is incremented past the 4-bit resolution it goes to 0000 in gray code, preserving the 1-bit difference in the wraparound.

Let's flip the script and say we just did a read on 14 and the read pointer is incremented to 15, when the write pointer is metastable at 15, but is really at 0. The buffer is therefore EMPTY, so further reads are prevented even though the data is there. Now the write pointer rolls over to 0, out of the metastable state.

read pointer: 1111   (1000 gray)
write pointer: 0000   (0000 gray)

Remember comparisons take place on the binary – so the read can go through on the last written value at location 15. The read pointer advances to 0, and now, since the pointers are equal and the cause is a read, the buffer is EMPTY again.

Now let's say the read has been lagging and the write has cycled all the way around to position 3 and the write pointer is incremented to 4. The read pointer is also 4, but it's metastable. So the buffer is FULL (equal pointers after a *write* cause FULL status).

read pointer: 0100   (0110 gray)
write pointer: 0100   (0110 gray)

Now the gray-code bit flips over so the binary value of the read pointer is 5.

read pointer: 0101   (0111 gray)
write pointer: 0100   (0110 gray)

The values aren't equal and so the write is allowed at position 4.

19.6. Footprint

The thesis by Ryan Apperson [5] describes a dual-clock FIFO that works at the correct speeds for the cube. It uses an SRAM as the core memory with a dimension of 16 bits (word size) x 32 entries. This totals 512 bits. The display cube FIFO requires 72 bits (header / data) x 32 levels (per group) x 4 groups = 9216 bits. This is 18 times more memory than the Apperson FIFO. Apperson used 180 nm technology, which is already extremely small, but the design was created in 2002. Process technology is now even smaller; as of 2021 it's down to 7 nm. The Apperson FIFO, though holding 18 times less memory than needed for the display cube application, fits exactly within the footprint of the base-plane silicon as described and illustrated at the end of Chapter 18. Space is at a premium but advances since 2002 include smaller process technology so that fitting 18 times more memory into the same footprint is feasible. The silicon for the Apperson FIFO is shown below.

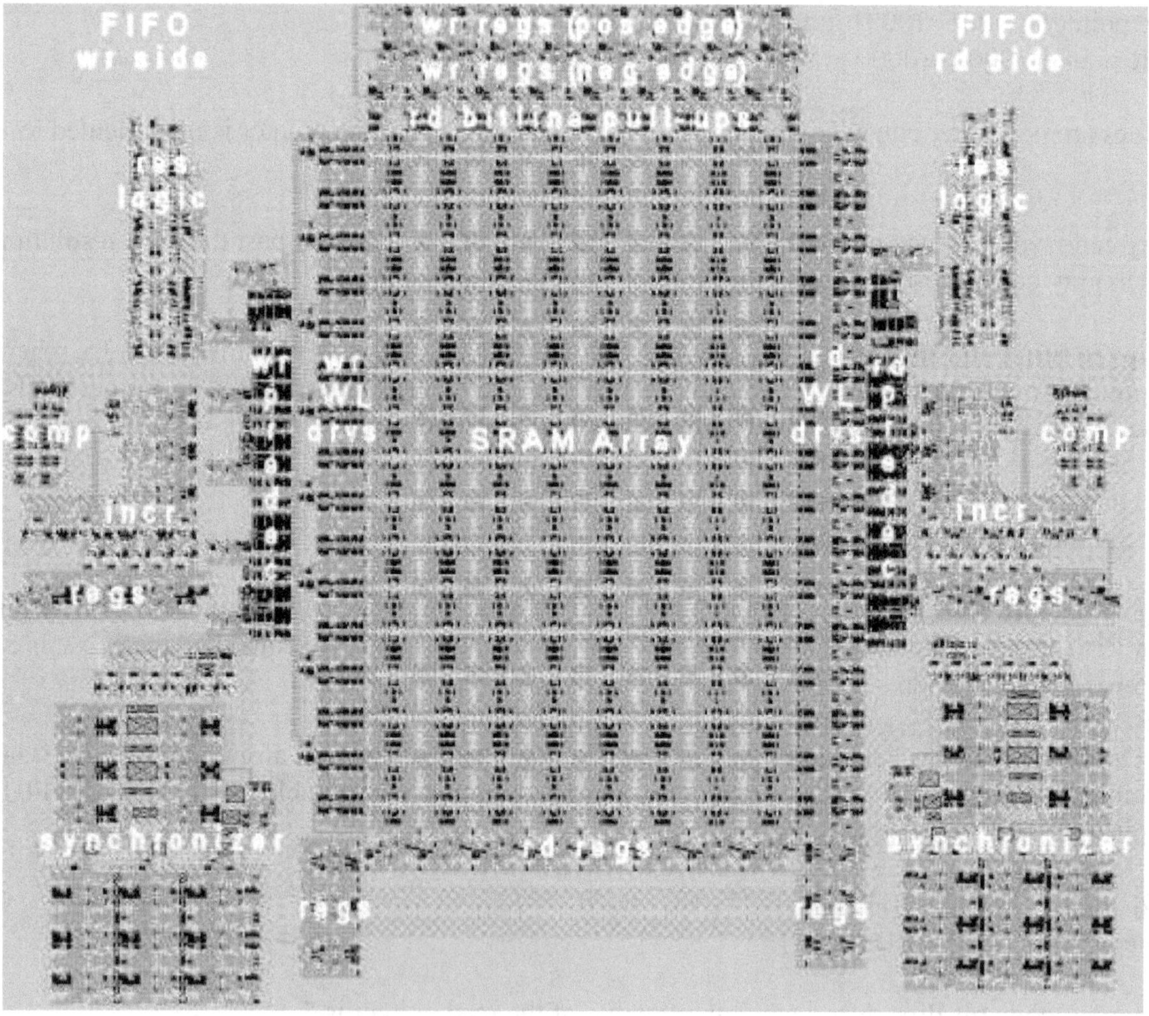

Figure 7. Apperson FIFO (Courtesy [5])

The label "res logic" stands for *reserve logic*. The FIFO has an *In Reserve* signal that goes high when the FIFO can't accept any more data. The FIFO keeps track of a nominal *reserve space* to handle data already in transit between the producer and the consumer (the consumer being the FIFO). The reserve space is needed because in some designs there's latency in the handshake between the producer and the consumer. The reserve logic only exists on the write-side of the FIFO, where the high-speed data is arriving. The reserve logic does some pointer math to determine that the FIFO isn't quite FULL but is *almost full*. This generates a FULL signal until the number of free slots exceeds the nominal reserve space. This is all well and good but since the cube application receives a set number of bytes, tailored to the total size of each FIFO, the almost full signal isn't important. The important thing is an *almost empty* signal, which generates a request to send (RTS) that brings on the next burst of data.

19.7. Conclusion

This chapter has given a taste of the complexity of the critical FIFO design for the display cube. Existing designs show that a dual-clock high-speed FIFO is not only feasible but a practical reality. This book's Appendix contains a Very High Speed Integrated Circuit (VHSIC) Hardware Description Language (VHDL) specification for the signal design of the FIFO and associated DRAM. It's not a final specification, just a hint of what's possible.

References

[1] Nebhrajani, Vijay. *Asynchronous FIFO Architectures, Parts 1, 2, and 3*. Available online (three separate downloads from different sites).
[2] Ramesh, Kumar, and Reddy. *Asynchronous FIFO Design with Gray code Pointer for High Speed AMBA AHB Compliant Memory Controller*, IOSR-JVSP, v. 1, issue 3, 2012, online
[3] Pinkston and Duato. *Interconnection Networks, Computer Architecture: A Quantitative Approach*, 4th Edition, Appendix E, ppt online
[4] Alexandropoulos et al. *A novel 1.8 V, 1066 Mbps, DDR2, DFI-compatible, Memory Interface*, 2010 IEEE Annual Symposium on VLSI, online
[5] Apperson, Ryan W. *A Dual-Clock FIFO for the Reliable Transfer of High-Throughput Data Between Unrelated Clock Domains*, Thesis, UC Davis, 2002
[6] Mason, A. *Lecture notes 13.1 ECE 410*, Michigan State University, online

About the Author

Daniel Browning is a programmer, systems engineer, and writer living in the San Francisco Bay Area. When an undergraduate at UC Berkeley he studied computer architecture with David Patterson, inventor of the RISC chip. An erstwhile member of SIGARCH, he has retained his love of hardware design while putting his software engineering talents to work in Silicon Valley.

Daniel would like to thank Ed Berlin, cofounder of Cubicomp and patented inventor of the rotating 3D display device; and Robert Pariseau, coinventor of the Amiga computer and a department head at Cubicomp. At Cubicomp Ed and his investors saw an opportunity to do the impossible: create a custom framebuffer at an affordable price and provide animation and rendering software for the IBM PC. That was in 1985. During an extended lunch conversation in 1987, Robert Pariseau explained precisely why a 3D display device was impossible with 1987 technology. That conversation over Mexican food and beer fired the imagination and wasn't forgotten.

Appendix

VHDL Specifications for FIFO RAM

The FIFO code fragment comes from [1], and the DPRAM spec is constructed from a mix of online examples, of which there are many. The cube won't use FPGAs. The FIFOs will be custom VLSI at a very small scale, not mounted as separate components but integrated into the monolithic design of a single die. A full dual port memory can read and write on clock A, and read and write on clock B. However the cube only requires a fast write on clock A and a slower read on clock B. The fast clock will never be reading; and the slow clock will never be writing. Port A is therefore write only; Port B read only.

```
-- Dual port RAM backend
-- wena is write enable, rena is read enable
library ieee;
use ieee.std_logic_1164.all;
use ieee.std_logic_unsigned.all;

entity lib_ram_stdpram is
  generic (
    -- 10 bits address resolution, 16-bit words
    ADDR : natural := 10;
    DATA : natural := 16
  );
  port (
    a_clk  : in std_logic;
    b_clk  : in std_logic;

    a_adr  : in std_logic_vector (ADDR-1 downto 0);
    a_wena : in std_logic;
    a_wdat : in std_logic_vector (DATA-1 downto 0);

    b_adr  : in std_logic_vector (ADDR-1 downto 0);
    b_rena : in std_logic;
    b_rdat : out std_logic_vector (DATA-1 downto 0)
  );
end lib_ram_stdpram;

-- register transfer logic (rtl)
```

```
architecture rtl of lib_ram_stdpram is
  -- Shared memory
  type ram_type is array (0 to (2**ADDR)-1) of std_logic_vector (DATA-1 downto 0);
  shared variable ram : ram_type;

begin

   -- A port logic --

   write_A : process (a_clk)
   begin
      if rising_edge(a_clk) then
         if a_wena = '1' then
            ram(to_integer(unsigned(a_adr))) := a_wdat;
         end if;
      end if;
   end process;

   -- B port logic --

   read_B : process (b_clk)
   begin
      if rising_edge(b_clk) then
         if b_rena = '1' then
            b_rdat <= ram(to_integer(unsigned(b_adr)));
         end if;
      end if;
   end process;

end rtl;


-- FIFO logic

library IEEE, STD;
use IEEE.std_logic_1164.all;
use IEEE.std_logic_arith.all;
use IEEE.std_logic_unsigned.all;

entity status is
generic (
    ADDR : natural := 10
);
port (
   reset      : in std_logic;
   clk        : in std_logic;
   fifo_wr    : in std_logic;
   fifo_rd    : in std_logic;
```

```
    valid_rd     : out std_logic;
    valid_wr     : out std_logic;
    rd_ptr       : out std_logic_vector(ADDR-1 downto 0);
    wr_ptr       : out std_logic_vector(ADDR-1 downto 0);
    empty        : out std_logic;
    full         : out std_logic
);
end status;

architecture status_AB of status is
    signal rd_ptr_s      : std_logic_vector(ADDR-1 downto 0);
    signal wr_ptr_s      : std_logic_vector(ADDR-1 downto 0);
    signal valid_rd_s    : std_logic;
    signal valid_wr_s    : std_logic;

begin
    empty_P : process(clk, reset)
    begin
        if (reset = '1') then
            empty <= '1';
        elsif (clk'event and clk = '1') then
            if (fifo_wr = '1' and fifo_rd = '1') then
                -- do nothing
                null;
            elsif (fifo_wr = '1') then
                -- write unconditionally clears empty
                empty <= '0';
            elsif (fifo_rd = '1' and (wr_ptr_s = rd_ptr_s + '1')) then
                -- set empty
                empty <= '1';
            end if;
        end if;
    end process;

    full_P : process(clk, reset)
    begin
        if (reset = '1') then
            full <= '0';
        elsif (clk'event and clk = '1') then
            if (fifo_rd = '1' and fifo_wr = '1') then
                -- do nothing
                null;
            elsif (fifo_rd = '1') then
                -- read unconditionally clears full
                full <= '0';
            elsif (fifo_wr = '1' and (rd_ptr_s = wr_ptr_s + '1')) then
                -- set full
                full <= '1';
            end if;
```

```
      end if;
   end process;

   -- valid read and write signal logic
   valid_rd_s <= '1' when (empty = '0' and fifo_rd = '1');
   valid_wr_s <= '1' when (full = '0' and fifo_wr = '1');

   wr_ptr_s_P : process(clk, reset)
   begin
      if (reset = '1') then
         wr_ptr_s_P <= (others => '0');
      elsif (clk'event and clk = '1') then
         if (valid_wr_s = '1') then
            wr_ptr_s <= wr_ptr_s + '1';
         end if;
      end if;
   end process;

   rd_ptr_s_P : process(clk, reset)
   begin
      if (reset = '1') then
         rd_ptr_s_P <= (others => '0');
      elsif (clk'event and clk = '1') then
         if (valid_rd_s = '1') then
            rd_ptr_s <= rd_ptr_s + '1';
         end if;
      end if;
   end process;

   rd_ptr <= rd_ptr_s;
   wr_ptr <= wr_ptr_s;
   valid_rd <= valid_rd_s;
   valid_wr <= valid_wr_s;

   -- valid_wr/valid_rd to a_rena and b_wena resp.

end status_AB;
```

Index

Made in the USA
Monee, IL
02 November 2026

67384611R00136